T0134890

Springer Theses

Recognizing Outstanding Ph.D. Research

Aims and Scope

The series “Springer Theses” brings together a selection of the very best Ph.D. theses from around the world and across the physical sciences. Nominated and endorsed by two recognized specialists, each published volume has been selected for its scientific excellence and the high impact of its contents for the pertinent field of research. For greater accessibility to non-specialists, the published versions include an extended introduction, as well as a foreword by the student’s supervisor explaining the special relevance of the work for the field. As a whole, the series will provide a valuable resource both for newcomers to the research fields described, and for other scientists seeking detailed background information on special questions. Finally, it provides an accredited documentation of the valuable contributions made by today’s younger generation of scientists.

Theses are accepted into the series by invited nomination only and must fulfill all of the following criteria

- They must be written in good English.
- The topic should fall within the confines of Chemistry, Physics, Earth Sciences, Engineering and related interdisciplinary fields such as Materials, Nanoscience, Chemical Engineering, Complex Systems and Biophysics.
- The work reported in the thesis must represent a significant scientific advance.
- If the thesis includes previously published material, permission to reproduce this must be gained from the respective copyright holder.
- They must have been examined and passed during the 12 months prior to nomination.
- Each thesis should include a foreword by the supervisor outlining the significance of its content.
- The theses should have a clearly defined structure including an introduction accessible to scientists not expert in that particular field.

More information about this series at http://www.springer.com/series/8790

Mario Cabrera

Development of 15 Micron Cutoff Wavelength HgCdTe Detector Arrays for Astronomy

Doctoral Thesis accepted by the University of Rochester, USA

Mario Cabrera
Glenn Dale
MD, USA

ISSN 2190-5053 ISSN 2190-5061 (electronic)
Springer Theses
ISBN 978-3-030-54243-6 ISBN 978-3-030-54241-2 (eBook)
https://doi.org/10.1007/978-3-030-54241-2

This Springer imprint is published by the registered company Springer Nature Switzerland AG
The registered company address is: Gewerbestrasse 11, 6330 Cham, Switzerland

To my wife Michelle, my son Julian, my mom Gabi, and my dad Mario.

Supervisor's Foreword

Mario Cabrera has been an outstanding member of our Infrared Instrumentation group at the University of Rochester. It has been my pleasure to serve as one of Mario's supervisors during his tenure as a PhD student, and we are proud that he continues to work on unique infrared detector arrays for astronomical space missions in his present position. As one of his PhD supervisors (with Prof. William Forrest), I am extremely pleased to introduce Mario's thesis work. For many years, the ternary $Hg_{1-x}Cd_xTe$ alloy has figured as a favored material for infrared detector arrays because of its tunable bandgap through modification of the Cd molar concentration, enabling production of short wavelength cutoff devices to very long wavelength cutoffs. While substantial progress had been demonstrated for short cutoff wavelength (2.5 μm) and midwave cutoff wavelength (5 μm) astronomical detector arrays, until our group tackled the problems inherent in 10 μm devices for the proposed NASA mission NEOSM (formerly known as NEOCam), they were not appropriate for the low astronomical radiation backgrounds of space. Mario tackled the much more difficult problem of extending the cutoff wavelengths of astronomical HgCdTe arrays to 13 and 15+ μm—where the Hg concentration of the alloy is much higher, and hence the material is much softer and much more prone to defects leading to high dark current at the defect positions on the array. Our group has traditionally worked with industry, providing characterization of produced devices to further illuminate changes in direction on the next steps to the reduction of high dark current pixels. For this process, Mario modeled the I–T and I–V characteristics of first 10 μm devices, directing our industrial collaborator on requirements for the manufacture of 13 μm, and finally 15+ μm devices. Building on the prevalence of dominant dark current mechanisms as a function of bias, focal plane temperature, and other fundamental physical properties of the material, we achieved substantial success in some of the few lot splits we ordered. In addition, our recommended architectural changes in array production were implemented, as well as lessons learned along the way in the production of successful short and midwave devices. Consequently, there has been interest expressed by other groups in the technology that Mario has pushed forward, and we expect even more interest as these devices are utilized. These longer wavelength arrays can be used in well-designed passively cooled space missions, or in space missions requiring modest cooling, in particular exoplanet missions designed to search for evidence of biosignature features, including in the important 15 μm CO_2 line. In addition, ground-based astronomical observations can benefit from this technology, and various groups are making inquiry. I am delighted that Dr. Cabrera's thesis work is being recognized by Springer, which will bring this important work to an even wider audience.

Professor Emerita of Astronomy
University of Rochester
Rochester, NY, USA

Judith L. Pipher

Acknowledgements

This work was supervised by a dissertation committee consisting of Professors William Forrest (co-advisor), Judith Pipher (co-advisor), and Dan Watson of the Department of Physics and Astronomy, Professor Gary Wicks of The Institute of Optics, and Professor Miki Nakajima of the Department of Earth and Environmental Sciences.

The detectors used in this work were provided by Teledyne Imaging Sensors (TIS), and the author has had many discussions regarding the technology with TIS scientists Dr. Donald Lee, Dr. Jianmei Pan, and Dr. Majid Zandian.

The graduate work was supported by NASA grant NNX14AD32G S07, New York Space grant, and the Graduate Assistance in Areas of National Need (GAANN) grant.

Special recognition is given to my advisors Professor William Forrest and Professor Judith Pipher and to the Senior Research Engineer Craig McMurtry of the UR infrared detector group. It has been a privilege to work with such a talented group of scientists. I am grateful to Bill for being a great mentor and advisor. The discussions we have had over the past 4 years have helped me grow as a scientist. All of the stories he has shared have had a teaching moment, and who better to learn about the history of infrared technology for astronomy than from Bill and Judy. I will forever be indebted to Judy for her guidance, feedback, and encouragement as an advisor. Judy has been a great role model and all of her knowledge and enthusiasm for all things astronomy and detector physics has been invaluable in my development as a graduate student. I am thankful to Craig for sharing all of this technical expertise and without him all the data presented here would have taken much longer to gather. I will miss working with the three of you.

I would also like to thank my lab mates and fellow grad students Meghan Dorn and Greg Zengilowski for all of their assistance in the lab and the great times I shared with them.

My sincere thanks also go to Professor Dan Watson for his encouragement, advice, and creating a welcoming environment in the department. I also thank Laura Blumkin, Rich Sarkis, Dave Munson, and Mike Culver who have all been very supportive.

Most importantly, I would want to acknowledge the four most important people in my life: mom, dad, Michelle, and Julian. This would not have been possible without the encouragement, support, and love of my mom and dad. Their belief in me always pushed me forward in uncertain times, and I would not be where I am today if it was not for them. Mom, I miss you dearly, and I wish you were here.

Words cannot express my appreciation for all the joy my wife Michelle has brought into my life these past 10 years, and I cannot imagine having a better partner and best friend to share all of this with. Julian, you beat me to it buddy! You came into our lives as I was finishing the fourth chapter. Your smiles and hugs helped me cope with the stress of finishing the last three chapters. You both mean everything to me, and I love you very much.

Parts of This Thesis Have Been Published in the Following Journal Articles

- McMurtry, C.W.; Cabrera, M.S.; Dorn, M.L.; Pipher, J.L.; Forrest, W.J.; "13 micron cutoff HgCdTe detector arrays for space and ground-based astronomy", Proc. SPIE **9915**, 99150E (2016).
- Cabrera, M. S., McMurtry, C. W., Dorn, M. L., Forrest, W. J., Pipher, J. L., and Lee, D., "Development of 13 μm Cutoff HgCdTe Detector Arrays for Astronomy," Journal of Astronomical Telescopes, Instruments, and Systems **5**(3), 1–18 (2019).
- Cabrera, M.S.; McMurtry, C.W.; Forrest, W.J.; Pipher, J.L.; Dorn, M.L.; Lee, D.; "Characterization of a 15 μm Cutoff HgCdTe Detector Array for Astronomy," Journal of Astronomical Telescopes, Instruments, and Systems **6**(1), 1–9 (2019).

Contents

List of Symbols

A	Diode junction area
C_0	Nodal capacitance at zero-collected signal
C	Pixel capacitance
c	Speed of light
D_h	Hole diffusion coefficient
E	Electric field across the depletion region
E_C	Conduction band energy
E_D	Donor atom energy level
E_F	Fermi energy
E_g	Bandgap energy
E_i	Intrinsic Fermi energy
E_t	Energy of traps that contribute to trap-to-band tunneling current
$E_{t_{gr}}$	Energy of traps that contribute to G-R current
E_V	Valence band energy
G_{ex}	Electron–hole pair generation from external sources
$G_{external}$	External array controller gain
G_{mux}	Multiplexer gain
h	Planck's constant
$\hbar$	Reduced Planck's constant
$I_{ideal\ dark\ current}$	Dark current of an ideal diode
I_0	Saturation current for an ideal diode
$I_{band\text{-}to\text{-}band}$	Band-to-band quantum tunneling dark current
I_{diff}	Diffusion dark current
$I_{G\text{-}R}$	Generation–recombination dark current
$I_{trap\text{-}to\text{-}band}$	Trap-to-band quantum tunneling dark current
$I_{photocurrent}$	Photocurrent
$\vec{k}$	Electron wavevector
k_b	Boltzmann's constant
L_h	Minority-carrier diffusion length

M	Transition matrix element
m_{eff}	Electron effective mass
n	Electron concentration in conduction band
n_{n_0}	Majority carrier concentration at thermal equilibrium
Δn	Excess majority carrier concentration
n_1	Electron density in conduction band when Fermi level falls at trap energy in depletion region
N_C	Conduction band effective state density
N_D	Donor atom density (doping density on n-type material)
N_D^+	Ionized donor atom density
n_i	Intrinsic carrier concentration
n_t	Trap density which contributes to trap-to-band tunneling current
n_{t_i}	Initial active trap density
n_{t_d}	Activated trap density
N_V	Valence band effective state density
p	Hole concentration in valence band
p_{n_0}	Minority carrier concentration at thermal equilibrium
Δp	Excess minority carrier concentration
p_1	Hole density in valence band when Fermi level falls at trap energy in depletion region
$\vec{p}$	Momentum
q	Charge of an electron
$S_{Background}$	Background signal
$S_{Center\ Output}$	Measured output signal in a pixel
$S_{Center\ Actual}$	Actual signal in a pixel (not affected by IPC)
$S_{Neighbor\ Output}$	Measured output signal in a neighboring pixel
$S_{[e]}$	Collected charges in a diode
$S_{[V]}$	Output signal in volts
$S_{output,\ [ADU]}$	Output signal in ADU
S_{input}	Input referred signal
S_{output}	Output referred signal
S_p	Surface recombination velocity
T	Temperature
U	Net rate of recombination in depletion region
$V_{0,\ actual}$	Initial actual bias across the diode
$V_{applied\ bias}$	Applied bias across the diode
V_a	Trap activation voltage
V_{bi}	Built-in bias across a $p-n$ junction
$V_{dark\ current}$	Change in bias due to dark current
$V_{detector\ bias}$	Bias across the diode
V_{Dsub}	Substrate applied voltage
$V_{n,\ actual}$	Actual voltage for nth SUTR sample
$V_{n,\ sample}$	Output voltage for nth SUTR sample
$V_{pedestal\ injection}$	Change in bias due to pedestal injection

V_{reset}	Reset voltage
V_{ZBP}	Zero bias point voltage
W	Depletion region width
x	Mole fraction of cadmium in HgCdTe
α	Interpixel capacitance coupling parameter
β	Band-to-band tunneling current fitting parameter
Γ	Symmetry point in reciprocal lattice
γ	Scaling factor
ϵ_0	Permittivity of free space
ϵ	Relative permittivity of HgCdTe
η	Quantum efficiency
θ	Anti-clockwise angle from the horizontal to the perpendicular of a line
θ_α	Angle between cross-hatching lines parallel to the $\left[\overline{2}31\right]$ and $\left[\overline{2}13\right]$ directions
θ_β	Angle between cross-hatching lines parallel to the $\left[\overline{2}31\right]$ and $\left[01\overline{1}\right]$ directions
λ_c	Cutoff wavelength
ρ	Perpendicular distance from the origin to a line
$\sigma_{Total,\ [e]}$	Total noise from collected charges
$\sigma_{Shot,\ [e]}$	Shot noise from collected charges
$\sigma_{System,\ [e]}$	System noise in electrons
$\sigma_{[V]}$	Noise in output signal voltage
$\sigma_{output,\ [ADU]}$	Noise in output signal in ADU
$\sigma_{IPC\ corrected,\ [ADU]}$	IPC corrected noise in output signal in ADU
τ_h	Minority carrier lifetime
τ_{n_0}	Electron lifetime in depletion region
τ_{p_0}	Hole lifetime in depletion region
Φ	Photon flux
ϕ_n	Electron quasi-Fermi electrostatic potential
ϕ_p	Hole quasi-Fermi electrostatic potential
ψ	Electrostatic potential

Chapter 1
Introduction

1.1 Astronomical Motivation

Space-based missions using infrared astronomy technologies have revolutionized the field of astronomy, increasing our knowledge of the evolution of the universe and our own solar system. Instruments in past space missions that used long wave infrared (LWIR) detector arrays (wavelength cutoff above $\sim$5 μm) required cooling to very low temperatures with on-board cryogens. For example, the Spitzer Space Telescope's Si:As impurity band conduction (IBC) LWIR arrays with a cutoff wavelength of $\sim$28 μm used in all three instruments (IRAC, MIPC, IRS) [1, 2] were operated at 6–8 K, and WISE's similar arrays (centered at 12 and 22 μm) [3] were cooled to 7.8 K. After cryogens ran out, due to good thermal design the focal plane of Spitzer Space Telescope warmed up and equilibrated to $\sim$27.5 K. The mid-wave IR InSb cameras continue to function at comparable sensitivity to that during the cryogenic phase, while the Si:As IBC LWIR large format detector arrays ceased to function due to high dark currents. Similarly, only the WISE mid-IR Mercury Cadmium Telluride (HgCdTe) arrays continued to function once cryogen was depleted. LWIR detector arrays that are cooled via cryogens or cryo-coolers on space missions take up valuable space and weight, and the limited volume of cryogens limits the lifetime of the LWIR detector array sensitivity.

The University of Rochester infrared detector group has worked together with Teledyne Imaging Sensors (TIS) for many years to address these limitations by developing HgCdTe detector arrays with a cutoff wavelength of $\sim$10 μm (LW10 arrays) for future low-background passively cooled missions. The success of those arrays (see Sect. 1.5) led to the interest in extending the cutoff wavelength of these devices further, and determine whether this material is a viable option to cover a larger range of the infrared spectrum without the need for cryogens, or with less stringent operating temperatures than Si:As detector arrays. This technology already exists for high-background applications, but would not be adequate for low-background astronomical observations.

M. Cabrera, *Development of 15 Micron Cutoff Wavelength HgCdTe Detector Arrays for Astronomy*, Springer Theses, https://doi.org/10.1007/978-3-030-54241-2_1

This thesis project focuses on the development of detector arrays that can operate at relatively elevated focal plane temperatures for low-background astronomy with a cutoff wavelength of 15 μm. TIS manufactures and develops these LWIR detector arrays with substantial feedback from our group. The testing and characterization of these devices is carried out at the University of Rochester (UR).

The target cutoff wavelength goal of 15 μm (LW15 arrays) for this project was chosen for future missions aimed at studying the atmospheres of exoplanets. Those missions would benefit from these detector arrays because solar systems with planets (or forming planets) can be detected with far better contrast at infrared rather than visible wavelengths. This technology would also enable the detection of CO_2 at 15 μm, a signature indicative of a terrestrial planet in the habitable zone [4]. This CO_2 feature, along with the detection of bio-signatures such as O_3 and H_2O would be crucial in identifying candidate exoplanets that may support life. In addition to space missions, this technology can be applied for ground based observatories. Instruments developed to explore the N-band, would be able to operate with the use of a cryo-cooler, eliminating the need for cryogens.

To extend the cutoff wavelength to 15 μm, the first step of this project was to first develop HgCdTe detector arrays with a cutoff wavelength of ∼13 μm (LW13 arrays). This step was crucial to identify any roadblocks that would prevent the project from reaching the final goal of developing 15 μm cutoff wavelength HgCdTe arrays.

1.2 Mercury Cadmium Telluride (HgCdTe)

Mercury Cadmium Telluride is an alloy composed of equal parts of Hg and Cd from column II in the periodic table to Te from column VI, resulting in the mixed crystal $Hg_{1-x}Cd_xTe$, where x is the composition parameter (molar fraction of Cd). The closely packed atoms in any crystal have overlapping discrete energy levels. As a consequence of the Pauli exclusion principle, no two electrons may occupy the same energy level, causing the discrete energy levels from individual atoms to split and form a continuum of energy levels or energy bands. The allowed electron energies in the crystal are functions of the electron wavevector $\vec{k}$, where the wavevector is related to the momentum carried by the electron wave $\vec{p} = \hbar\vec{k}$.

In semiconductors and insulators, the energy bands are arranged in such a way that there is a range of energies where no electron energy levels exist (assuming a perfect crystal without dislocations or impurities). This gap in allowed energy states is referred to as the band gap, and it results from the interaction between electrons and the periodic potential from the ion cores where a solution to the electron wavefunction does not exist [5, 6]. The band gap energy is defined as the difference between the highest energy of the valence band and the lowest energy of the conduction band. The valence band is formed by the overlapping of neighboring valence (bonding) p orbitals, while the other energy states due to antibonding are elevated to higher energy values, where the lowest of these elevated levels (s orbitals) form the conduction band. The bonding p-orbitals form three distinct valence bands: light hole, heavy hole, and the spin-orbit energy bands. The spin-orbit energy band has a much lower energy than the light hole and heavy hole band, both of which are degenerate at $k = 0$. The key difference between insulators and semiconductors is if for a given operating temperature, there is enough thermal energy in the system to excite electrons to the conduction band. The concentration of electrons in a solid that are thermally excited to conductive states is proportional to $\exp\left(-\frac{E_g}{2k_bT}\right)$ [6, 7] for an intrinsic semiconductor. If the band gap energy is too large, the valence band would remain completely filled, where no electrons would be capable of conducting. At a temperature of absolute zero, all semiconductors act as an insulator since all of the filled energy states lie below the valence band, while at higher temperatures there is a non-vanishing probability of thermally exciting electrons to the conduction band.

The desirability of $HgCdTe$ as an infrared detector array is the ability to tune the band gap energy by varying the Hg to Cd ratio in the alloy, where the band gap energy (in eV) at a temperature in degrees Kelvin is given by [7]

$$E_g(x,T) = -0.302+1.93x-0.81x^2+0.832x^3+5.35\times10^{-4}T(1-2x), \qquad (1.1)$$

while the relationship between the band gap energy and the cutoff wavelength is given by

$$E_g = \frac{hc}{\lambda_c}, \qquad (1.2)$$

where h is Planck's constant and c is the speed of light.

HgTe ($x = 0$) is a semimetal for which the conduction band has an energy lower than the valence band (inverted band order), whereas the semiconductor CdTe ($x = 1$) has a band gap energy of ~1.6 eV at a temperature of 80 K [8, 9] (1.5 eV at 300 K). The crossover in composition for $Hg_{1-x}Cd_xTe$ where this material turns from a semimetal to a semiconductor happens at $x \approx 0.16$, where the band gap is ~0 eV.

In E-k space, the highest valence band energy (Γ_6) and the lowest conduction band energy (Γ_8) in HgCdTe occurs at $k = 0$ (Γ symmetry point in the reciprocal lattice or k-space), making HgCdTe a direct gap semiconductor. The energy *vs.* k diagram for HgCdTe near $k = 0$ is shown in Fig. 1.1, where the band gap energy is $E_g = \Gamma_6 - \Gamma_8$. For values of $x < 0.16$, the band gap energy $\Gamma_6 - \Gamma_8$ is negative.

1.3 Photo-Diodes

The band gap in semiconductors allows for the fabrication of photon absorbing devices, where the band gap energy sets the lower limit on the photon energies

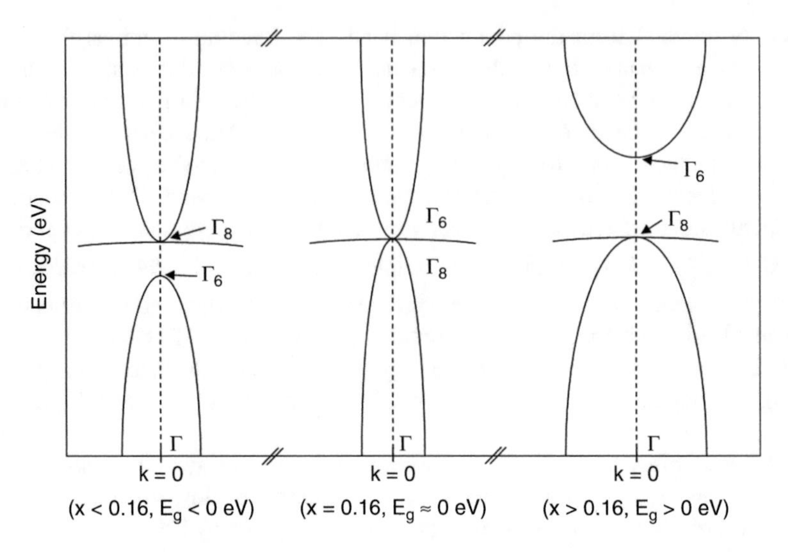

Fig. 1.1 Energy *vs.* k diagram for $Hg_{1-x}Cd_xTe$ for differing composition values of x

that can be absorbed by the material. When photons are absorbed, an electron-hole[1] pair is created, where the electron transitions from the valence band to the conduction band and photo current flows through the material. The conducting electrons and holes have random trajectories within the material and can recombine, losing information about the absorbed photons. To avoid recombination of the electrons and holes, the detector arrays presented here have a p-n heterojunction structure, where an electric field near the junction separates the electrons and holes.

To properly understand the operation and characteristics of these devices, it is important to first present important conduction properties of semiconductor materials, and p-n junction characteristics.

1.3.1 Intrinsic and Extrinsic Semiconductor Properties

Semiconductor devices are categorized as either intrinsic or extrinsic. Semiconductors are considered intrinsic if the material has no added impurities, while an extrinsic semiconductor has added impurities with energy levels within the forbidden energy gap of the intrinsic semiconductor. In extrinsic n-type material, atoms having one more valence electron than is needed to form a covalent bond in the crystal are added (while still maintaining charge neutrality), and are referred to as donors, where the act of adding these atoms is called doping. The excess

[1]Holes correspond to empty energy levels, and can act as a positive charge carriers with charge $+q$ under the influence of electric and magnetic fields.

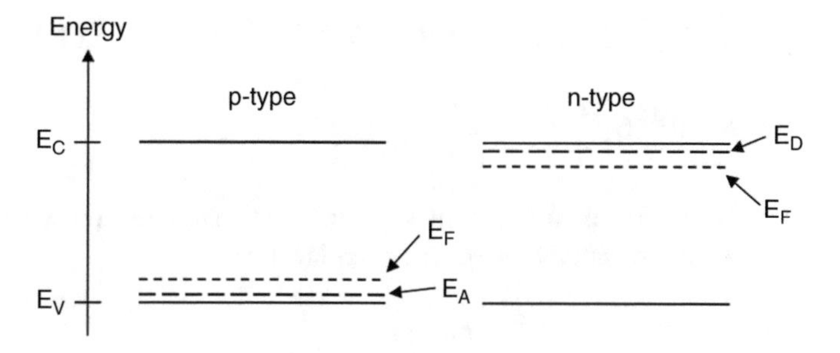

Fig. 1.2 Energy diagram for extrinsic semiconductors. The p-type semiconductor shows the acceptor energy levels E_A near the valence band E_V, and the n-type semiconductor shows the donor energy levels E_D near the conduction band E_C. The Fermi level is shown for both extrinsic materials as E_F

electrons are weakly bound, and create available impurity energy levels close to the conduction band. The added atoms to make p-type material have one fewer valence electron than is needed to form a covalent bond. The atoms with missing electrons are referred to as acceptors, and form available impurity energy levels near the valence band of the material. The band diagram for both p- and n-type material are shown in Fig. 1.2, where the Fermi level is closer to the conduction band and the valence band for n- and p-type material respectively.

The Fermi level is defined as the energy level required to maintain charge neutrality in the material. For an intrinsic material in thermal equilibrium, this means that the electron concentration in the conduction band n is equal to the hole concentration in the valence band p. The charge neutrality condition is therefore presented as

$$n = p = n_i, \tag{1.3}$$

where n_i is the intrinsic carrier concentration. The electron concentration in the conduction band and hole concentration in the valence band are given by

$$n = N_C \exp\left[\frac{E_F - E_C}{k_b T}\right] \tag{1.4}$$

$$p = N_V \exp\left[\frac{-(E_F - E_V)}{k_b T}\right], \tag{1.5}$$

where k_b is the Boltzmann constant, E_F is the Fermi energy, E_C and E_V are the conduction and valence band energies respectively, N_C and N_V are the effective density of states in the conduction and valence band respectively, and the intrinsic carrier concentration for HgCdTe it is estimated to be [7]

$$n_i = \left(5.585 - 3.820x + 1.753 \times 10^{-3}T - 1.364 \times 10^{-3}xT\right) \times \left[10^{14} E_g^{3/4} T^{3/2} exp\left(-\frac{E_g}{2k_b T}\right)\right], \tag{1.6}$$

where x is the cadmium mole fraction. In addition to the charge neutrality condition, the product np must also satisfy the mass-action law [10, 11]

$$np = n_i^2. \tag{1.7}$$

This product is independent of the Fermi level, and must be satisfied in intrinsic and extrinsic (doped) materials.

In an n-doped material, at a temperature $T > 0$, there is a non-zero probability that the extra electrons from the donor atoms will transition to the conduction band due to the small difference in energy levels, leaving a net $+q$ charge at the site of the donor atom (the donor atom was neutral before being ionized), making electrons the majority carriers and holes are the minority carriers. For the acceptor atoms (p-type), the close proximity of the impurity energy level to the valence band allows for electrons in the valence band to fill the acceptor energy level to complete the covalent bond, leaving a hole in the valence band. The ionized acceptor level will have a $-q$ charge at the site of the acceptor atom. In p-type material, holes are the majority carriers, while electrons are the minority carriers.

To determine the Fermi level in an extrinsic material, the ionized impurities need to be considered in the charge neutrality condition. For the n-type material, the electron density in the conduction band must equal the hole density in the valence band plus the density of ionized donor atoms [10, 11]

$$n = N_D^+ + p, \tag{1.8}$$

where n and p are given by Eqs. 1.4 and 1.5, and the number of ionized atoms N_D^+ is [11]

$$N_D^+ = N_D\left[1 - F\left(E_D\right)\right] = N_D \frac{\exp\left[\left(E_D - E_F\right)/k_b T\right]}{1 + \exp\left[\left(E_D - E_F\right)/k_b T\right]}, \tag{1.9}$$

In Eq. 1.9, N_D is the doping density of donor atoms, $F\left(E_D\right)$ is the Fermi-Dirac function evaluated at the donor energy level E_D. The Fermi-Dirac function is the probability of having an electron occupying the energy level E_D, therefore the quantity $\left[1 - F\left(E_D\right)\right]$ is the probability of not having an electron in the donor energy level (i.e. probability of an ionized donor energy level). Rewriting the charge neutrality equation, we obtain

$$N_C \exp\left[\frac{E_F - E_C}{k_b T}\right] = N_D \frac{\exp\left[\left(E_D - E_F\right)/k_b T\right]}{1 + \exp\left[\left(E_D - E_F\right)/k_b T\right]} + N_V \exp\left[\frac{-\left(E_F - E_V\right)}{k_b T}\right], \tag{1.10}$$

where the Fermi level is determined so that the equality is true.

1.3.2 p-n Junction

A p-n junction is formed by bringing an n-type and p-type material together. Once in contact, the excess electrons from the n-type material diffuse towards the p-type material to fill the empty covalent shells in the p-type material. The electrons diffusing towards the p-type material can simultaneously be interpreted as holes diffusing from the p- to n-type material. Both the n- and p-type material were neutral before coming into contact, this electron/hole diffusion in the junction creates an electric field pointing from the n-type material to the p-type material since there is a positively and negatively (respectively) net charge distribution near the junction. This is a self-limiting mechanism since the strength of the electric field will prevent further diffusion, occurring when the Fermi levels of the p- and n-type material are the same as shown in Fig. 1.3. The region in which all covalent bonds have been completed or do not have excess electrons needed to complete the bonds is called the depletion (or space-charge) region.

The voltage across the depletion region due to the self-limiting diffusion of majority carriers is called the built-in bias and is given by [12]

$$V_{bi} = \frac{k_b T}{q} \ln \frac{N_A N_D}{n_i^2}, \tag{1.11}$$

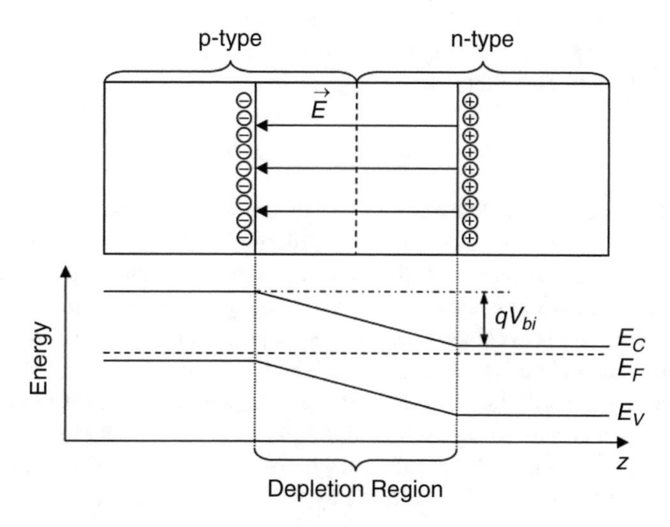

Fig. 1.3 p-n junction diagram with zero applied bias

where N_A is the acceptor doping density. The built-in bias can also be approximated by [10, 13]

$$V_{bi} \approx \frac{E_g}{q}. \tag{1.12}$$

In p-n diodes, there are thermal and tunneling effects that can give rise to currents under thermal equilibrium. These currents, not due to the absorption of light, are called dark currents and the theory for these currents are presented in Chap. 3. For an ideal diode, without traps or impurities, the dark current is given by the ideal diode equation

$$I_{ideal\ dark\ current} = I_0 \left(\exp\left[\frac{q V_{detector\ bias}}{k_b T} \right] - 1 \right), \tag{1.13}$$

where I_0 is the magnitude of the dark current,[2] and $V_{detector\ bias}$ is the voltage (or bias) across the diode. When the diode is operated under the forward bias regime ($V_{detector\ bias} > 0$), the polarity of the applied bias opposes the built-in bias, making the diode conductive, leading to an exponentially increasing dark current. If a reverse bias ($V_{detector\ bias} < 0$) is applied, the ideal dark current is small and quickly becomes very stable since the exponent term in Eq. 1.13 approaches zero.[3] Figure 1.4 shows the current versus bias behavior of this ideal current. Increasing the reverse bias further can result in junction breakdown, where due to tunneling effects (discussed in Sect. 3.3), and avalanche multiplication [10] the diode will conduct very large currents. Figure 1.4 also shows the current *vs.* bias behavior under illumination. The photocurrent is given by

$$I_{photocurrent} = \eta q \Phi, \tag{1.14}$$

where η is the quantum efficiency of the diode and Φ is the photon flux (photons/sec/pixel) incident on the diode. Under illumination two features of I-V curves must be discussed: the short circuit current I_{SC}, and the open circuit voltage V_{OC}. The short circuit current occurs when the net voltage across the diode is zero. The short circuit current is exclusively due to photocurrent since the dark current is zero when there is no net voltage on the diode. The I-V curve of a reverse biased diode under illumination will enter the forward bias regime at this short circuit current point. The open circuit voltage refers to the point where the net current across the junction is zero. At this open circuit voltage, the diode is said to be saturated since photocurrent can no longer be collected by the depletion region. It is important to

[2] The details of this parameter and derivation of this dark current component (diffusion) is presented in Sect. 3.1.

[3] In the detector arrays presented here, the exponent term goes to zero with ~25 mV of reverse detector bias, where the typical applied reverse bias ranges from 50–350 mV.

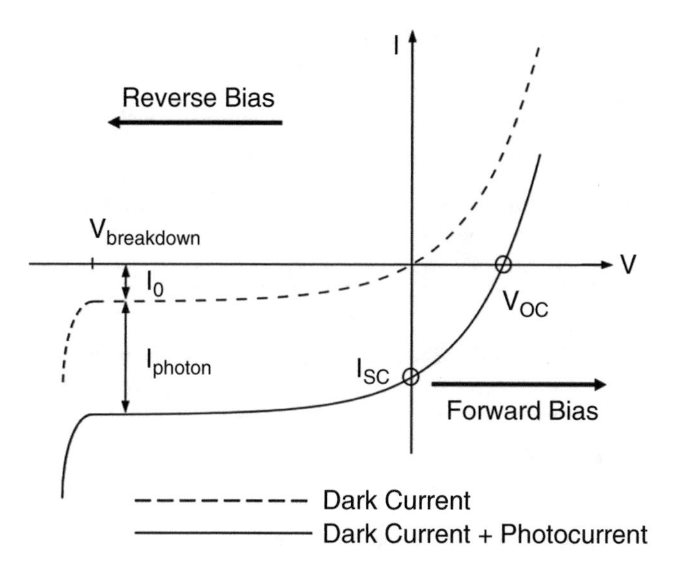

Fig. 1.4 Dark current *vs.* bias for an ideal photo-diode

note that the saturation point reached when the diode is illuminated occurs under forward bias. If the diode is not illuminated, the dark current can saturate the array. This saturation point occurs when the applied bias is zero, and the net current across the p-n junction is zero. This saturation point, in the absence of illumination, is referred to as the zero bias point V_{ZBP}.

1.3.3 Photo-Diode Structure

Figure 1.5 shows the structure of individual pixels in the arrays that are presented here. The n-type HgCdTe bulk material is grown on a nearly lattice matched CdZnTe substrate using molecular-beam epitaxy (MBE) growth technique. In MBE growth, the material (Hg, Cd, and Te) is evaporated and deposited on the substrate, allowing for single layer growth under high vacuum. Innovations by TIS and the UR using this growth technique has resulted in lower dislocation densities, which have consequently reduced dislocation dependent dark currents. p-type HgCdTe is then implanted into the n-type material to form individual photo-diodes. The photo-diodes are then bonded with indium bumps to the circuitry that enables operation of the arrays.

These arrays are back-side illuminated on the bulk n-type material (photons initially passing through the substrate), and absorbed photons in the bulk HgCdTe material will create an electron-hole pair. The holes that diffuse to the p-n junction will then be separated from the majority carriers (electrons in the n-type material) by the electric field in the depletion region, and will be collected in the p-type

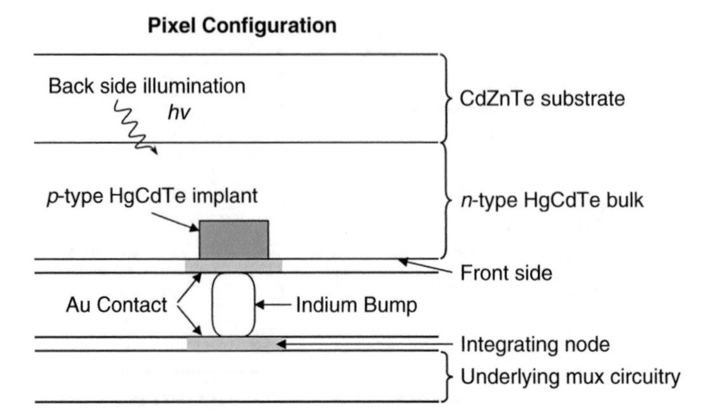

Fig. 1.5 Pixel structure (not to scale)

implant. The collection of holes in the p-type material then changes the potential at the integrating node, where the integrating node is connected to the p-type implant by an indium bump, allowing us to measure the amount of holes that were collected.

When the applied reverse bias is increased, the depletion region will increase by further diffusing electrons from the n-type material towards the p-type material, and vice versa with holes from the p-type material diffusing to the n-type side of the junction, enhancing the electric field in the depletion region. The collected charges from dark current and/or photo-current on the p-side of the junction will oppose the polarity of the electric field, decreasing the depletion region width (debiasing) until saturation is reached and the net current through the junction is zero. The main advantage in increasing the reverse applied bias is the increase of the depletion region width, which will allow more charges to be collected thus increasing the well depth.[4] The drawback of increasing the reverse bias is a potential surge of dark current in non-perfect arrays. The dark current bias dependence will be further discussed in Chap. 3, and the effects will be shown in Chaps. 5 and 6.

1.4 Multiplexer

Following the fabrication of the photo-diodes, to operate the device, the array of photo-diodes must be bonded to a readout integrated circuit (ROIC), also referred to as multiplexer. Fowler [14] showed that sampling the array n times a short time after the reset, followed by n more times an integration time later can reduce the noise in the image by $\sqrt{n}$. The final image using Fowler-sampling is formed by averaging the first n frames, and subtracting from the average of the final n frames.

[4]Well depth is the measure of how much charge the diode can collect before it reaches saturation.

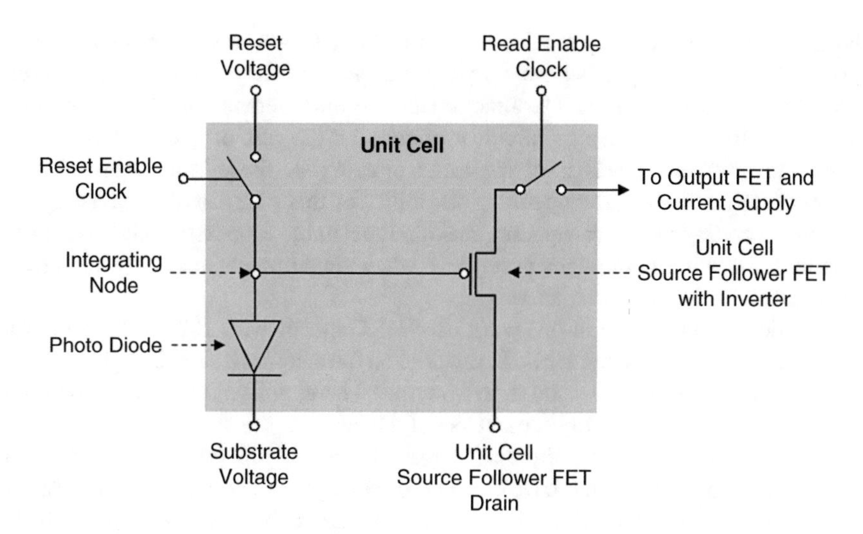

Fig. 1.6 Direct Readout unit cell schematic

The details of this method are further discussed in Sect. 2.4.2. The ability to read out the array multiple times as the photo-diode debiases is also beneficial to study different characteristics such as the dark current and the non-linear capacitance.

Direct Readouts (DRO) are non-destructive readouts that use Complementary Metal Oxide Semiconductor (CMOS) technology, where each individual pixel in the array can be addressed separately, without the need of transferring charges from one pixel to another. Instead, the voltage at the integrating node, due to the collection of charge, acts as the gate voltage for the source-follower field-effect transistor (FET) in the unit cell circuitry for each individual pixel (shown in Fig. 1.6). When specific pixels are read out, the switch connecting the source-follower FET to the output is closed, and remains open when the pixel is not being addressed. This type of readout provides the ability to read out the pixels non-destructively.

1.4.1 HAWAII-1RG Multiplexer

The HAWAII-XRG (**H**gCdTe **A**stronomy **W**ide **A**rea **I**nfrared **I**mager with **X**k×Xk resolution, **R**eference pixels and **G**uide mode) is a multiplexer (or mux) that uses the CMOS technology to read out detector arrays non-destructively. As detector arrays for astronomy grow in size to provide a larger field view, while at the same time decreasing the size of individual pixels to increase the resolution of images, the circuitry required to read out the array non-destructively is quite complex. The HAWAII-XRG (HXRG from here on) has multiple operating modes such as providing multiple outputs to enhance the read out data rate, the ability to read out only certain portions of the array, and it also provides eight rows (top-most and bottom-most four) and eight columns (left-most and right-most four) of reference pixels. The reference pixels have a capacitor in place of the light-sensitive material used for the detector array. The functionality of the reference pixels is to correct for any voltage drifts due to circuitry effects for the rest of pixels in the array since the reference pixels have the same circuitry as those with light-sensitive material. In addition to the operating flexibility of this mux, another advantage is the low operating power dissipation, making it desirable for passively cooled space missions. Drawing current per pixel only when single pixels are being addressed leads to the low power dissipation.

The devices presented in this work use H1RG muxes with 1024×1024 pixels, with the long-wave infrared HgCdTe material on the central 1016×1016 pixels. Two outputs are used to read out the arrays presented here, with a full array integration time of ~5.8 s, and a full array reset time of 5.8 ms.

For some photo-diodes, the single reset before signal integration can be a drawback. Photo-diodes that have constant or negligible dark currents with respect to changing bias can be well calibrated since the dark current will remain relatively constant throughout the integration of signal. For photo-diodes that exhibit large bias-dependent dark currents, can be difficult to calibrate since the dark current will change throughout the integration of signal following the single reset. Tunneling currents are an example of dark currents that vary exponentially with changing bias. The longer wavelength cutoff of the LW13 and LW15 devices are susceptible to these non-linear tunneling dark currents due to relatively smaller band gap, compared to the 5 μm devices developed for JWST or the LW10 devices for the NEOCam mission.

1.4.2 CTIA Multiplexer

Multiplexers based on the capacitive transimpedance amplifier (CTIA) structure, which have a feedback capacitor, can maintain the detector bias relatively constant as charge is being integrated [15]. The advantage of a multiplexer based on this structure is that the feedback capacitor can be large enough to increase the pixel well depth by a considerable amount with respect to the HXRG mux, with a small and constant reverse detector bias. An integration of charge with this type of mux would maintain a constant dark current, in addition to having the flexibility to operate at a bias where the dark current is small. As a reminder, a larger reverse applied bias is needed for HXRG muxes if the user requires larger well depths, whereas a CTIA mux would not. The main drawback of conventional CTIA mux is that it draws a much larger current to operate due to the continuous feedback process to maintain the detector bias for all pixels. The larger power dissipation from this mux would not be ideal for passively cooled space missions. The proposed NEOCam mission requires 1 mW/megapixel or less to achieve sufficient passive cooling.

1.5 Prior Development

The UR infrared detector team has been working on the development and improvement of low background devices with a cutoff wavelength longer than about 5.4 μm since 1992 [16–21]. The first deliveries of HgCdTe detectors that were delivered to UR by Rockwell (company name at the time) were single photo-diodes [17] wire bonded to a readout. These best performing diodes with cutoff wavelength of 10.6 μm (at a temperature of 40 K) had dark currents $>10^3$ e^-/s at temperatures between 20 and 40 K with an applied reverse bias of 20 mV. As part of the work presented in Wu (1997 PhD thesis) [17], five HgCdTe photo-diodes with a cutoff wavelength of 13.7 μm (at 30 K), showed dark currents $>10^6$ e^-/s at temperatures between 20 and 40 K with an applied reverse bias of 20 mV.

By 2003, early large format arrays were bonded to an H1RG mux with a pixel pitch of 36 μm. Each pixel was bump bonded to every other unit cell in the H1RG mux, rendering an effective 512 × 512 pixel array. Two of these first large format arrays had cutoff wavelengths of 9.3 and 10.3 μm [22], showed low dark current, but could not support a large reverse bias, which limited the well depth of these devices. Three other arrays with cutoff wavelengths ranging from 8.4 to 9.1 μm, also bump bonded to an H1RG mux with an effective size of 512 × 512 pixels, showed dark currents below 30 e^-/s and well depths of at least 40 mV at a temperature of ~30 K for more than 70% of pixels. Even with these limitations, the devices showed promise as they were operable at a temperature of 30 K.

1.5.1 Near Earth Object Camera

The Near Earth Object Camera (NEOCam) is a passively cooled NASA Jet Propulsion Laboratory (JPL) proposed mission whose goal is to find and characterize near-Earth objects (asteroids and comets), some of which have Earth-crossing orbits and could be potentially hazardous. The mission will have two cameras, NC1 and NC2. NC1 will cover the wavelength range from 4–5 μm using HgCdTe detector arrays with a cutoff wavelength of ~5 μm (already developed for WISE and the James Webb Space Telescope [23]), while NC2 will use ~10 μm (LW10) HgCdTe detector arrays to allow the telescope to observe from 6–10 μm.

To meet the requirements of the NEOCam project, TIS successfully increased the LW10 array format from the 512 × 512 pixel arrays presented in Bacon [20] to 2048 × 2048 pixels with an 18 μm pitch [24]. The arrays have cutoff wavelengths of ~10 μm, low read noise, low dark current, high quantum efficiency, and have operabilities >90%[5] up to temperatures of 42 K, and have been proton irradiated

[5]In addition to high QE and low read noise, the operability requirements also include well depths of more than 46,000 e^- and dark current less than 200 e^-/s at 40 K, which will be the focal plane temperature for NC2. Since most pixels have shown to have good QE and read noise, much of the

to demonstrate the ability of these arrays to withstand the cosmic ray radiation they are expected to receive in space for the detector lifetime [24–26]. The dark current operability requirement of 200 e^-/s for NEOCam's NC2 was established such that it will be background-limited by the thermal emission from the zodiacal dust cloud. One of the biggest improvements for these arrays was the ability to apply larger biases of up to 350 mV and maintain low dark currents for the majority of pixels. To meet the well depth requirement for NEOCam, 250 mV of detector reverse bias is needed.

1.6 Extending the Cutoff Wavelength to 15 μm

The successful improvement in dark current and the ability to apply larger reverse biases to obtain larger well depths in LW10 arrays, encouraged the UR detector team to pursue a project with the goal of extending the cutoff wavelength of these devices to determine if this is a viable material option for this longer wavelength range over the current Si:As detector arrays.

In order to reach the present 15 μm goal ($x \sim 0.209$, at a temperature of 30 K) of this project, an intermediate step of developing arrays with a cutoff wavelength of 13 μm ($x \sim 0.216$, at a temperature of 30 K) was taken to identify any problems that would prevent us from reaching the 15 μm goal.

As the compound becomes softer (which is a consequence of increasing the mercury fraction with increased wavelength), there is an increased likelihood of defects/dislocations [27] that would contribute to trap-assisted tunneling currents. The smaller band gap also leads to an increase in direct band-to-band tunneling.

The LW10 arrays have shown excellent QE (>70% before anti-reflection coating), satisfactory image quality, and low correlated double sample (CDS) read noise of 19 e^- [24]. High dark currents and/or low well depths are expected to limit the performance of the LW13 and LW15 arrays. For this developmental project, we have focused our efforts on the reduction of high dark currents since they are expected to worsen from the ~10 μm arrays when the wavelength is increased to 13 μm initially, and to the final goal of 15 μm.

1.6.1 Phase I: 13 μm Cutoff Devices

For the first phase of this project, four 1024 × 1024 pixel arrays bonded to H1RG multiplexers with a pixel pitch of 18 μm and a wavelength cutoff of ~13 μm were delivered from TIS to UR. Two of the four arrays, H1RG-18367 and H1RG-18508, were grown and processed in the same manner as the LW10 arrays for the proposed

development of the LW10 devices has been focused on reducing the dark currents at the required temperature of 40 K.

NEOCam mission, but extended to the desired 13 μm cutoff wavelength. The other two arrays, H1RG-18369 and H1RG-18509, use TIS/UR proprietary experimental designs to reduce tunneling dark currents.

The same dark current operability requirements of the LW10 devices developed for the NEOCam mission [24] are used to determine the best LW13 array design. Our imposed operability requirements include dark currents $<200\ e^-/s$, and well depth of $\sim 40\ ke^-$ for an applied bias of 150 mV (larger well depth requirements are used for larger applied biases). The operability requirements adopted for NEOCam are also specific to the filters, telescope size and throughput. Any future mission that intends to use these LW13 devices likely would have a modified set of requirements.

Though operability requirements depend on specific applications of these devices, the dark current and well depth requirements for operable pixels in the LW13 arrays are used as a benchmark to compare the performance of the different arrays at different temperatures and applied bias, and to determine the best pixel design that will be pursued when increasing the cutoff wavelength to 15 μm.

The characterization of the LW13 arrays, quickly revealed that tunneling currents are the primary dark current mechanisms limiting the operability of these devices at low temperatures and moderate high applied reverse bias, such as those used to characterize and meet the NEOCam requirements for the LW10 arrays.

In addition to tunneling dark currents, thermal (diffusion [12] and generation-recombination [28]) dark currents have also been found to be sources of dark current in these LW13 devices. Tunneling dark currents (band-to-band and trap-to-band) are strong functions of bias, and have large non-linear effects that can affect data calibration. Well depth of pixels with very large dark currents, due to trap-to-band tunneling, can be diminished leading to low well depths. A subset of these pixels with large dark currents and/or low well depth create a cross-hatching pattern in the operability map for all four arrays (see Sect. 5.1.5), which has been identified to form due to the crystal axis mismatch between the CdZnTe substrate and the detecting HgCdTe layers [29, 30]. This cross-hatching pattern has also been observed in LW10 devices, though much less prominent and those arrays have higher operabilities.

Although a measurable contribution from tunneling dark currents on operable pixels was measured at larger biases, the proprietary experimental design of H1RG-18509 successfully lowered tunneling dark currents in comparison to the other three devices. The median dark current and well depth at a temperature of 28 K and reverse applied bias of 350 mV for this array is 1.8 e^-/s and 81 ke^- (385 mV[6]) respectively, while the median dark current for the other three arrays was $>200\ e^-/s$. The effects of tunneling currents on calibration and operability for all four devices are shown in Sects. 5.1.1 and 5.1.4.

The dark current of these devices has been modeled using the theory described in Chap. 3, with results of the fitted models to data presented in Sect. 5.2. At biases $>\sim 200$ mV and low temperatures (28 K), it has been shown that band-to-band

[6]This larger well depth than the applied bias is due to the pedestal injection contribution, see Sect. 2.3

tunneling is the dominant component of dark current and it is quite uniform across the arrays since all pixels for a given device are affected equally by this dark current mechanism.

1.6.2 Phase II: 15 μm Cutoff Devices

The diode structure of H1RG-18509 was selected as the best structure to move forward with the production of ∼15 μm cutoff wavelength detector arrays. The three arrays that we received for the second phase of the project from TIS: H1RG-20302, 20303, and 20304 have cutoff wavelengths of 16.7, 15.5, and 15.2 μm respectively at 30 K. The analysis of the three LW15 arrays showed an improvement in the diode structure to reduce tunneling currents further. For H1RG-20303, at a temperature of 23 K and an applied bias of 350 mV, the measured dark current is approximately two orders of magnitude smaller of what was expected from an array with the exact same performance as LW13 array H1RG-18509, and extending the cutoff wavelength to 15.5 μm.

The longer cutoff wavelength of these LW15 devices leads to larger dark currents if operated at similar temperatures and biases as the LW13 devices. Figure 1.7 shows the median dark current *vs.* temperature for all LW13 and LW15 devices

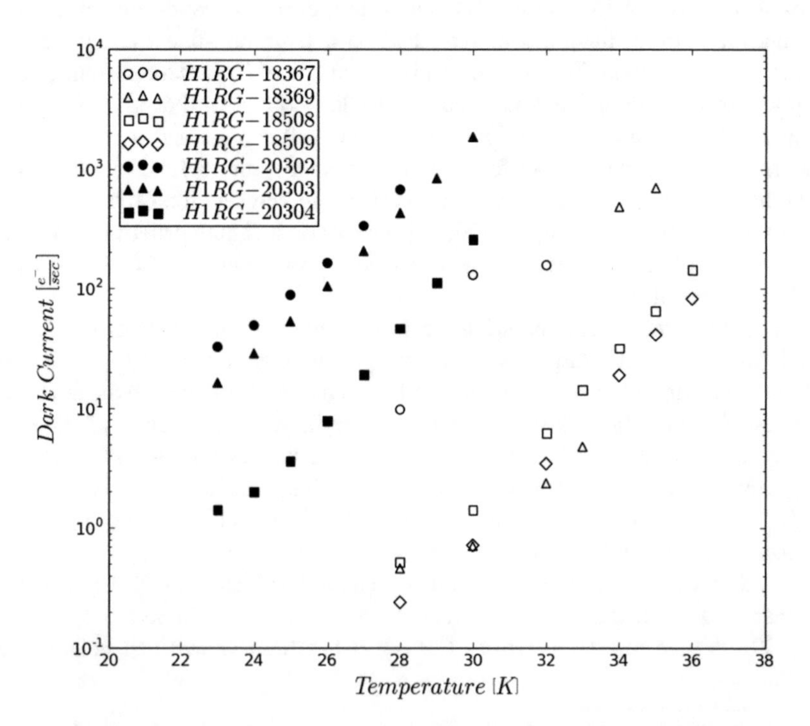

Fig. 1.7 Median dark current *vs.* temperature for all LW13 and LW15 devices with an applied bias of 150 mV. The solid data points correspond to the LW15 arrays

with an applied bias of 150 mV. The LW15 devices show dark currents ~2–3 orders of magnitude higher than the LW13 devices at a temperature of 28 K. It should be pointed out that the "dark current" measured for LW13 device H1RG-18367 included a glow from the mux at all three temperatures. This elevated current was uniform across the entire array, and is not consistent with any of the dark current mechanisms studied here. H1RG-18369 also showed this uniform mux glow in dark current measurements at temperatures of 34 and 35 K with this applied bias (this glow was also present in dark current measurements with 250 mV of reverse bias and read noise data for this device). The increase by a factor of ~100 in current from 33 to 34 K is not due to dark current mechanisms. This mux glow was only observed for these two devices, and will be further discussed in Sects. 5.1.2, 5.1.4, and 5.2.2.

Although the LW15 devices would require to be operated at lower temperatures to achieve comparable dark currents to the LW13 devices, at 28 K, the shortest wavelength LW15 device had a median dark current below 100 e^-/s. These results are very encouraging since the thermal dark current in LW15 devices at these temperatures appears to be dominated by G-R dark currents which depend on trap sites in the depletion region. With further improvement on the processing method of these arrays, the dark current may continue to improve for future devices. Dark current results for the LW15 devices are presented in Sect. 6.1.3, and the dark current model for these devices is shown in Sect. 6.2.

Chapter 2
Test and Data Acquisition Setup

Before describing the data that is required to characterize the detector arrays and calibrate the data, a summary of the test setup and operation of the H1RG multiplexer is presented in this chapter.

2.1 Camera Dewar Setup

UR uses a low-background liquid helium camera dewar with multiple chambers that houses the detector arrays. The innermost chamber of the dewar (housing the array and filter wheel) has an aluminum cylindrical shield coupled to a liquid helium (LHe) reservoir, allowing temperatures down to 4 K and below to be reached. To achieve temperatures below 4 K (boiling temperature of LHe), the pressure in the LHe reservoir needs to be reduced below ambient pressure. The dewar layout is shown in Fig. 2.1.

The filter wheel has several narrow band filters with central wavelengths commonly used for astronomical observations, two circular variable filters (covering a wavelength range between 4–14.3 μm), and a dark blocking filter (or cold dark shutter). The housing of the filter wheel is made of blackened aluminum, with a small aperture attached underneath (facing outward towards the warmer shields) used to narrow the illumination pattern, which will pass through the filter in place. The initial aperture is referred to as the "Lyot helper". On the opposite side of the Lyot helper on the filterwheel is a 67.6 μm diameter Lyot stop used to control the illumination of the array. When used as a camera at a telescope, the telescope's secondary mirror is imaged onto the Lyot stop to minimize the thermal background while the stars are imaged on the array itself.

The inner chamber is surrounded by an aluminum cylindrical shield attached to a liquid nitrogen reservoir, shielding the liquid helium reservoir and inner chamber from the room temperature outer shell, extending the lifetime of the liquid helium.

M. Cabrera, *Development of 15 Micron Cutoff Wavelength HgCdTe Detector Arrays for Astronomy*, Springer Theses, https://doi.org/10.1007/978-3-030-54241-2_2

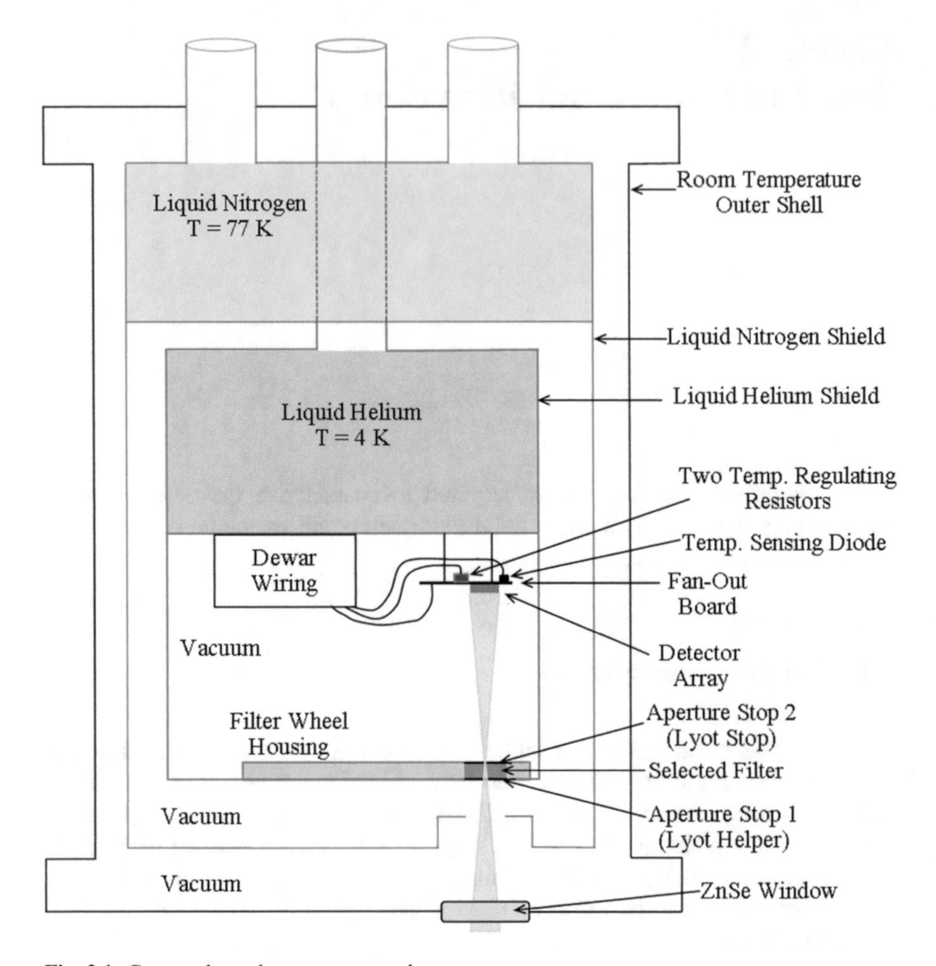

Fig. 2.1 Camera dewar layout, not to scale

The liquid nitrogen has an opening to allow illumination to reach the filter wheel, and will not define the solid angle of the beam. The outer shell has an anti-reflection coated ZnSe window in the line of sight of the array being tested with a transmission >80% from ~2.6–14 μm.

The detector array is connected to a fan-out board with a specific scramble cable provided by TIS. The fan-out board is mounted to the liquid helium reservoir with brass posts. In order to thermally insulate the fan-out board from the liquid helium reservoir, nylon nuts and washers were used, allowing us to operate at temperatures above 4 K. The temperature inside the innermost chamber is regulated with a Lake Shore Cryotronics temperature controller, allowing the arrays presented here to be tested at different stable temperatures. A temperature sensing diode is attached to the fan-out board on the opposite side of the detector array, where two resistors (encapsulated in copper to reduce glow) are used to regulate the temperature inside the innermost chamber. The fan-out and temperature controlling system is connected to the dewar wiring, emerging from the inner chamber with airtight sealed connectors to the top of the dewar (near the cryogen reservoir necks), which then connects to the external array controller.

2.2 Array Controller

The array controller used for data acquisition is based on an open source hardware design developed by the Observatory of the Carnegie Institute of Washington (OCIW). The version that is used to obtain all of the data presented here is optimized for infrared arrays.

The external array controller consists of a boxed backplane with four slots that contain the power supply, digital signal processing (DSP) board, clocking card, and a digitizer. The DSP used is a Motorola DSP56303, and it provides the clock sequence patterns to the clocking card. The clocking card then provides the biases in the order needed by the multiplexer to control the array. The 16-bit analog to digital (A/D) converters in the video card then converts the analog output of the multiplexer into digital form. The video card has been modified to handle four output channels from the multiplexer (though the H1RG can be programmable to use up to 16 outputs), where only two outputs are used to read out the data from the H1RG multiplexer.

The output signal from the multiplexer is amplified with a gain of 10.98 before it is digitized by the video card. The array signal frames are then saved in Flexible Image Transport System (FITS) format and sent to the host computer. The digital signal is stored in analog to digital units (ADUs), which can be converted to volts by dividing the 5 V range of the 16 bit A/D converter by 2^{16} ADUs and the gain from our array controller electronics.

In addition to providing an interface between the host computer and the detector array, the array controller also monitors the temperature diode voltage used inside of the dewar to control the array temperature, and controls the motion of the filter wheel set by the user through the host computer.

2.3 H1RG Multiplexer Operation

The H1RG mux is operated by using two of the 16 available output channels, with a pixel read out rate of 100 kHz. Resetting the array is done row-by-row, where all pixels in a row are reset simultaneously. During the reset stage, when a pixel row is selected, the reset enable clock in the unit cell of the individual pixels in the row will connect the reset voltage V_{reset} to the integrating node. The difference in voltage between the integrating node and the voltage applied to the substrate V_{Dsub} is the applied bias $V_{applied\ bias}$. In reverse bias, $V_{Dsub} > V_{reset}$, opposing the polarity of the built-in bias. Once the reset enable clock disconnects the integrating node from the reset voltage, the diode will begin to debias due to the collection of charges produced from absorbed photons and/or dark current. It is important to note that the actual bias immediately following reset does not necessarily equal to the applied bias. Large dark currents can debias the pixels between reset and the first time the pixel is read out, leading to an initial actual bias lower than the applied bias. After the reset switch is turned off to allow the device to debias, a redistribution of charge (pedestal injection) due to capacitive coupling to the reset FET results in an additional change in the bias voltage across the diode, which can typically increase the initial actual bias across the diode $V_{0,\,actual}$ by 0–75 mV[20, 22]. The initial actual bias is then given by

$$V_{0,\,actual} = V_{applied\ bias} - V_{dark\ current} + V_{pedestal\ injection}. \tag{2.1}$$

This initial actual bias is also called the diode well depth since this is the voltage (due to accumulated charge) that the photo-diode would need to debias to reach saturation.

After each pixel in the array (or sub-array) has been reset, when a pixel's unit cell is selected individually to read out the voltage from the collected charge in the photo-diode, the read enable clock connects the unit cell's source-follower (UCSF) FET to the output FET. The voltage at the integrating node acts as the gate voltage of the UCSF FET, allowing current to flow from the drain to the output FET. The voltage at the integrating node is amplified by both the UCSF and output FET. The total multiplexer gain is designed to be near unity, and this gain is measured to calibrate the data (see Sect. 4.2.1).

This method of reading out the referred charge collected in the p-type implant, allows us to make several measurements as the diode debiases, and measure I-V characteristics of the photo-diode. The current is given by the ratio of the collected charge and the time it took to collect. To determine the actual voltage at any given sample n sharing the same reset, the zero-bias point (V_{ZBP}) must be known, and is therefore important to measure the saturation level in the dark. The actual voltage for a sample n would then be given by

$$V_{n,\,actual} = V_{n,\,sample} - V_{ZBP}. \tag{2.2}$$

If the saturation level is measured when the diode is under illumination (V_{OC}), and used as a reference voltage to determine the actual voltage on the photo-diode, there will be a forward bias contribution to the actual reverse bias measurement (see Sect. 1.3.2). Section 4.4 will further discuss the method used for measuring V_{ZBP}.

2.4 Sampling Modes

2.4.1 Sample-Up-the-Ramp

Figure 2.2 shows a schematic of the SUTR method. Following the reset of the array (denoted as R in Fig. 2.2), the pedestal injection that occurs once the reset enable switch opens is represented by the discontinuous drop in output signal between the reset and the start of the signal collection represented by the sloped line as a function of time. As it was mentioned in Sect. 1.4.1, the magnitude of this pedestal injection varies from reset to reset, and can contribute to larger well depths. Following the reset and pedestal injection, the first sampled image is referred to as the pedestal (P). The delay time to the pedestal sample after reset ranges from circa 5.8 ms to 5.8 s, depending on the row and column for a specific pixel. This delay time will affect the well depth since signal will be collected before the pedestal is read out. The difference in signal between the reset and the pedestal frame here is the schematic equivalent to Eq. 2.1. After the pedestal, equally spaced frames in time are read out as signal is collected. The collection of samples that share the same reset is referred to as a ramp. The frame-to-frame integration time is what will be referred to as just "integration time" when describing the different data sets that are taken using this mode.

The SUTR method is used to obtain many non-destructive samples to study the behavior of the array as pixels debias over time. Data taken in SUTR mode also allows for easy cosmic ray hit correction since the deposited signal does not change over time and it is seen as a discontinuous increase in signal. The slope of the ramp before and after the cosmic ray hit, for a given affected pixel, is the same for linear integrated signal cases.

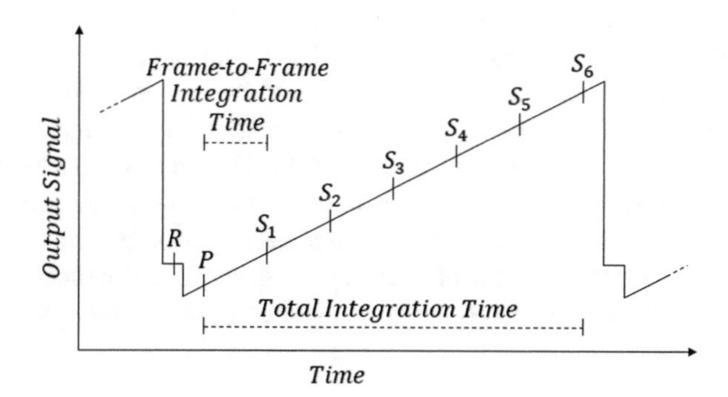

Fig. 2.2 SUTR readout mode diagram

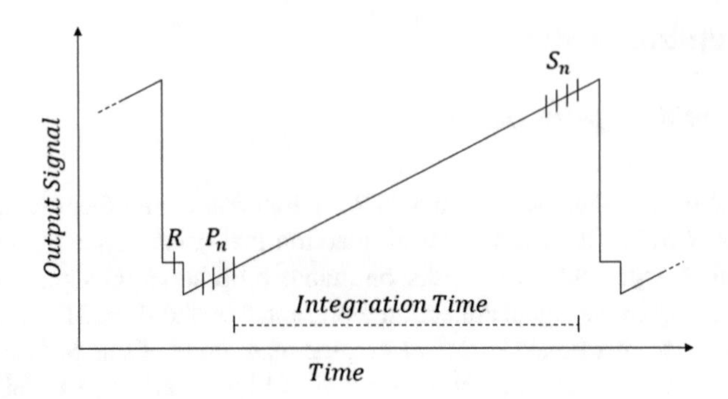

Fig. 2.3 Fowler-n sampling readout mode

2.4.2 CDS (Fowler-1)

Fowler sampling method [14, 31, 32] will be described, for completeness, since correlated double sample images are a special case of this sampling method.

Fowler and Gatley [14] described a multiple-read method to reduce the read noise in images taken with direct readout multiplexers. Their noise-reduction algorithm suggested reading n-samples immediately after reset, followed by n-samples an integration time later. A Fowler-n image is then formed by averaging the first n-samples (image pedestal frame), and subtracting it from the final n-samples (image signal frame). The integration time in this sampling method corresponds to the time between the nth pedestal frame and the nth signal frame. A schematic of this sampling method is shown in Fig. 2.3. Fowler and Gatley showed that this method reduces the noise in the final image by $\sqrt{n}$, only up to Fowler-32 images. Fowler sampling will only reduce the noise of the image if the system is dominated by the read noise, since for long enough integration times, $1/f$ noise would begin to have a larger contribution that cannot be mitigated by Fowler sampling.

A CDS image is equivalent to a Fowler-1 image, where only one pedestal frame is read out, and a signal image an integration time later. CDS images with different integration times can also be formed from an SUTR ramp, where the pedestal is subtracted from any of the signal frames in the ramp. CDS images are required to eliminate the KTC resetting Johnson noise associated with capacitive devices. CDS images are used when the noise is of interest such as in read noise measurements and capacitance measurements where the noise squared (due to photon flux) *vs.* signal method[33] is used.

Chapter 3
Dark Current Theory

Characterization of the LW10 devices [17, 20, 22] showed the presence of four main sources of dark current: diffusion, generation-recombination, band-to-band tunneling, and trap-to-band tunneling dark current. Bacon [20] showed effects of surface dark currents on the devices characterized for that work, but will not be covered here because improvements by TIS in the processing of their arrays have greatly reduced this dark current component, and have not been observed in any of these devices or in subsequent NEOCam LW10 devices. All of the dark currents presented here are described by mechanisms in units of electrons per second.

3.1 Diffusion Current

Diffusion dark current occurs when electrons in the valence band gain enough energy, thermally, to overcome the band gap and transition to the conduction band. The thermally generated electron-hole pair diffuse through the bulk material until they recombine if the electron-hole pair is not generated within one diffusion length of the space-charge region, or are separated by the electric field in the space-charge region.

The derivation of diffusion dark current is outlined in Reine et al. [12], and is closely followed here with the only difference that we treat the p-on-n junction photodiode case. The first assumption that is made here is that we have an abrupt p-n junction, allowing us to divide the diode into three regions with negligible transitions between the regions. These three regions are namely the n-type, space-charge (depletion), and p-type regions, shown in Fig. 3.1. Both n- and p-type regions are electrically neutral and are assumed to be uniformly doped.

Diffusion current of holes (minority carriers) in the n-type region is given by [10, 11]

M. Cabrera, *Development of 15 Micron Cutoff Wavelength HgCdTe Detector Arrays for Astronomy*, Springer Theses, https://doi.org/10.1007/978-3-030-54241-2_3

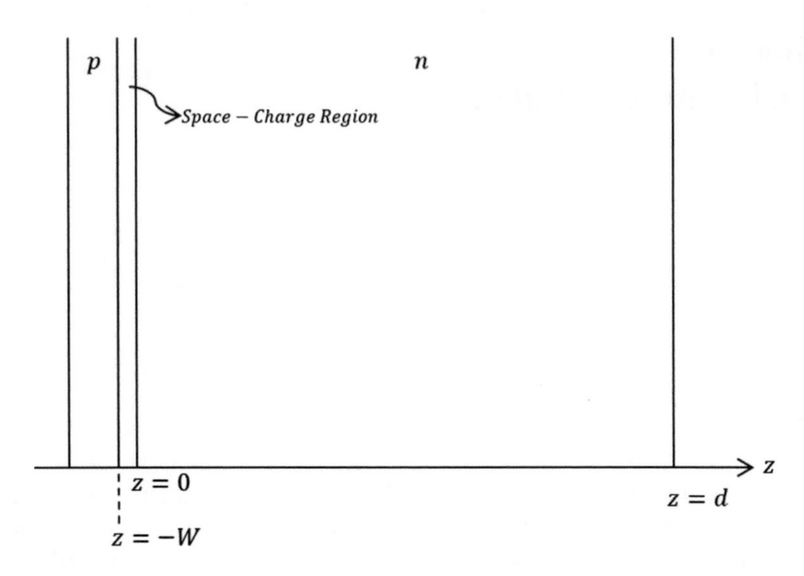

Fig. 3.1 Illustration of the p-n junction regions, not to scale

$$I_{diff} = -AD_h \frac{\partial p(z,t)}{\partial z}, \tag{3.1}$$

where A is the diode junction area, D_h is the hole diffusion coefficient, and $p(z, t)$ is the minority carrier concentration in the n region when the minority concentration at thermal equilibrium, p_{n_0}, is disturbed and given by

$$p(z,t) = p_{n_0} + \Delta p(z,t). \tag{3.2}$$

The majority carrier concentration is given by a similar expression, $n(z, t) = n_{n_0} + \Delta n(z, t)$, where due to the charge neutrality condition in the n region, $\Delta p(z, t) = \Delta n(z, t)$.

In thermal equilibrium, the carrier concentrations must also obey the mass-action law [10]

$$p_{p_0} n_{n_0} = n_i^2, \tag{3.3}$$

where n_{n_0} is equal to the doping density N_D in the n-type region, and n_i is the intrinsic carrier concentration (Eq. 1.6). The only unknown quantity needed to determine the diffusion current (I_{diff}) is the excess minority concentration Δp.

The minority carrier concentration $p(z, t)$ must follow the continuity condition in the n-type side [10, 11]:

$$\frac{\partial p}{\partial t} = G_{ex} - \frac{\Delta p}{\tau_h} - \frac{\partial J_{diff}}{\partial z}, \tag{3.4}$$

where G_{ex} is the generation of electron-hole pairs from external sources like photogeneration, τ_h is the minority-carrier lifetime, and J_{diff} is given by Eq. 3.1 divided by the diode junction area. The second term on the right of Eq. 3.4 corresponds to the rate at which excess holes are being thermally generated and recombined in the n-type material.

Assuming that there are no external sources contributing to the generation of excess minority carriers (i.e. $G_{ex} = 0$), the steady-state excess minority concentration must satisfy

$$0 = D_h \frac{\partial^2 \Delta p}{\partial z^2} - \frac{\Delta p}{\tau_h}. \tag{3.5}$$

The boundary conditions needed to solve Eq. 3.5 are:

$$p(0) = p_{n_0} exp\left(\frac{qV_{detector\ bias}}{k_b T}\right) \tag{3.6}$$

$$I_{diff}(z = d) = D_h \frac{\partial \Delta p}{\partial z}\bigg|_{z=d} = -S_p \Delta p(d). \tag{3.7}$$

Equation 3.6 is the hole density at the boundary of the space-charge and n-type region [10], while the second boundary condition (Eq. 3.7) is the diffusion current at the boundary of the n-type material and the substrate [12], where S_p is the surface recombination velocity.

The HgCdTe devices presented here are grown on a nearly lattice matched CdZnTe substrate by MBE, reducing the surface recombination velocity considerably ($S_p \approx 0$) compared to the diffusion velocity [20]. Under this assumption, and using both boundary conditions, a solution to the excess minority concentration from Eq. 3.5 is given by

$$\Delta p(z) = p_{n_0}\left[exp\left(\frac{qV_{detector\ bias}}{k_b T} - 1\right)\right]\frac{\cosh \frac{z-d}{L_h}}{\cosh \frac{d}{L_h}}, \tag{3.8}$$

where $L_h = \sqrt{D_h \tau_h}$ is the minority-carrier diffusion length. Now that we have Δp as a function of position z, we care about the diffusion current at the interface of the space-charge and n-type region ($z = 0$). Substituting Eq. 3.8 into 3.1, evaluating at $z = 0$, and taking the approximation of $d \ll L_h$ yields the final result

$$I_{diff} = A\frac{n_i^2 d}{N_D \tau_h}\left[exp\left(\frac{qV_{detector\ bias}}{k_b T}\right) - 1\right]. \tag{3.9}$$

In reverse bias, the detector bias across the diode ($V_{detector\ bias}$) is negative and the exponential term is negligible for the applied biases that we typically apply (>50 mV of reverse bias). It can be readily seen that diffusion dark current does not change appreciably with varying bias. The strong temperature dependence of this dark current is due to the squared dependence of the intrinsic carrier density.

3.2 Generation-Recombination

Generation-Recombination (G-R) dark currents are also a strong function of temperature where traps in the depletion region with energy levels between the valence band and the conduction band can facilitate the formation of electron-hole pairs through indirect transition of an electron to the conduction band (or conversely a hole to the valence band), where electrons in traps would have to overcome a smaller energy to become conductive. These generated electron-hole pairs are immediately separated by the electric field in the depletion region, reducing the recombination rates. The steady-state, nonequilibrium condition for the net rate of recombination is shown by Shockley and Read [34] to be

$$U = \frac{\left(pn - n_i^2\right)}{\left[\left(n + n_1\right)\tau_{p_0} + \left(p + p_1\right)\tau_{n_0}\right]}, \tag{3.10}$$

where τ_{p_0} and τ_{n_0} are the lifetimes for holes and electrons respectively in the depletion region, while p and p_1 are the density of holes in the valence band when the Fermi level falls in the intrinsic Fermi energy (E_i) and at the trap energy $E_{t_{gr}}$ respectively, and similarly for density of electrons in the conduction band n and n_1. Under nonequilibrium, the trap centers provide a net rate of generation, $U < 0$ (reverse bias), while $U > 0$ (forward bias) for net recombination [12].

Following the notation and method presented in Sah et al. [28], the relation of G-R current as a function of voltage is obtained by

$$I_{G-R} = A\int_{-W}^{0} U dz, \tag{3.11}$$

where A is the diode junction area, and the integration is over the depletion region of width W (see Fig. 3.1), where the depletion region width is given by [17]

$$W = \sqrt{\frac{2\epsilon\epsilon_0\left(V_{bi} - V_{detector\ bias}\right)}{qN_D}}, \tag{3.12}$$

where ϵ_0 is the permittivity of free space, ϵ is the relative permittivity of HgCdTe, and V_{bi} is the built-in voltage. To obtain U as a function of applied bias, the following relations

$$p_1 = n_i \exp\left[\frac{E_i - E_{t_{gr}}}{k_b T}\right] \tag{3.13}$$

$$n_1 = n_i \exp\left[\frac{E_{t_{gr}} - E_i}{k_b T}\right] \tag{3.14}$$

$$p = n_i \exp\left[\frac{(\phi_p - \psi)\, q}{k_b T}\right] \tag{3.15}$$

$$n = n_i \exp\left[\frac{(\psi - \phi_n)\, q}{k_b T}\right] \tag{3.16}$$

are substituted into Eq. 3.10, where ψ is the electrostatic potential, $\psi = -qE_i \approx -qE_g/2$, while ϕ_p and ϕ_n are the quasi-Fermi electrostatic potential for holes and electrons respectively. Making these substitutions into the net recombination rate gives the following expression:

$$U = \frac{n_i}{\sqrt{\tau_{p0}\tau_{n0}}} \cdot \frac{\sinh \frac{q(\phi_p - \phi_n)}{2k_b T}}{\cosh\left[\frac{q}{k_b T}\left(\psi - \frac{\phi_p + \phi_n}{2}\right) + \ln\sqrt{\frac{\tau_{p0}}{\tau_{n0}}}\right] + \exp\left[\frac{-q}{2k_b T}(\phi_p - \phi_n)\right]\cosh\left(\frac{E_{t_{gr}} - E_i}{k_b T} + \ln\sqrt{\frac{\tau_{p0}}{\tau_{n0}}}\right)}, \tag{3.17}$$

where $\phi_p - \phi_n = V_{detector\ bias}$. Assuming a linear potential variation across the depletion region, Sah et al. [28] make the approximation

$$\psi - \frac{\phi_p + \phi_n}{2} = \frac{V_{bi} - V_{detector\ bias}}{W}\left(z + \frac{W}{2}\right), \quad -W < z < 0. \tag{3.18}$$

Furthermore, making the simplifying assumption that the hole and electron lifetimes are equal, $\tau_{gr} = \tau_{p0} = \tau_{n0}$, and using the change of variable

$$y = \exp\left[\frac{q}{k_b T}\left(\frac{V_{bi} - V_{detector\ bias}}{W}\right)\left(z + \frac{W}{2}\right)\right], \tag{3.19}$$

G-R current will be given by

$$I_{G-R} = \frac{n_i W A}{\tau_{gr}}\left[\frac{sinh\left(\frac{-qV_{detector\ bias}}{2k_b T}\right)}{\frac{q(V_{bi} - V_{detector\ bias})}{2k_b T}}\right] f(b), \tag{3.20}$$

$$f(b) = \int_0^\infty \frac{dy}{y^2 + 2by + 1}, \tag{3.21}$$

$$b = exp\left[\frac{-qV_{detector\ bias}}{2k_b T}\right]\cosh\left(\frac{E_i - E_{t_{gr}}}{k_b T}\right). \tag{3.22}$$

G-R current, similarly to diffusion, also has a weak dependence on bias (as long as the back-bias is greater than ~50 mV). The strong temperature dependence of this dark current component is in the form of the intrinsic carrier density n_i. Although diffusion currents increase much faster with increasing temperature as n_i^2, at lower temperatures, G-R currents can have a considerable contribution to dark current over diffusion due to the shallower temperature dependence.

Additionally, at any given temperature, both diffusion and G-R will increase with increasing cutoff wavelength since thermally generated electron-hole pairs would require less energy to overcome the band gap. Therefore, to obtain similar dark currents due to thermally excited electron-hole pairs, longer wavelength cutoff arrays would have to be operated at lower temperatures.

3.3 Quantum Tunneling Currents

Tunneling currents are produced by electrons tunneling from the valence to the conduction band directly (band-to-band), or by traps with energies between the valence and the conduction band (trap-to-band). There are two different models for band-to-band currents; the first assumes a triangular barrier, while the second one uses a parabolic barrier. The equation corresponding to the parabolic barrier is omitted here as our data best matches the behavior of a triangular barrier.

Band-to-band tunneling current (also referred to as Zener breakdown) is given in Sze [10] by

$$I_{band\text{-}to\text{-}band} = -\frac{q^2 A E V_{detector\ bias}}{4\pi^2\hbar^2}\sqrt{\frac{2m_{eff}}{E_g}}exp\left(-\frac{4\sqrt{2m_{eff}}E_g^{3/2}}{3q\hbar E}\right), \tag{3.23}$$

where m_{eff} is the effective mass of an electron [13, 20] and E_g is the band gap energy. E is the peak electric field across the depletion region given by [12]

$$E = \sqrt{\frac{2N_D\left(E_g - qV_{detector\ bias}\right)}{\epsilon\epsilon_0}}. \tag{3.24}$$

Trap-assisted tunneling, as the name suggests, occurs when an electron transitions from the valence band to the conduction band via trap energy states. The optimum trap energy for this process, similar to G-R would be at the mid-band gap level if they are present. Additionally, the traps that contribute the most to trap-to-band tunneling may not necessarily be the same traps that contribute to G-R current. A general form of trap-to-band tunneling current for a triangular barrier is modeled by Kinch [13, 35] when the tunneling rate is the rate-limiting process by

$$I_{trap\text{-}to\text{-}band} = \frac{\pi^2 q m_{eff} A V_{detector\ bias} M^2 n_t}{h^3\left(E_g - E_t\right)}exp\left(-\frac{4\sqrt{2m_{eff}}\left(E_g - E_t\right)^{3/2}}{3qE\hbar}\right), \tag{3.25}$$

where E_t is the energy of the trap level with respect to the valence band, and n_t is the trap density in the depletion region at E_t, and M is the mass matrix associated with

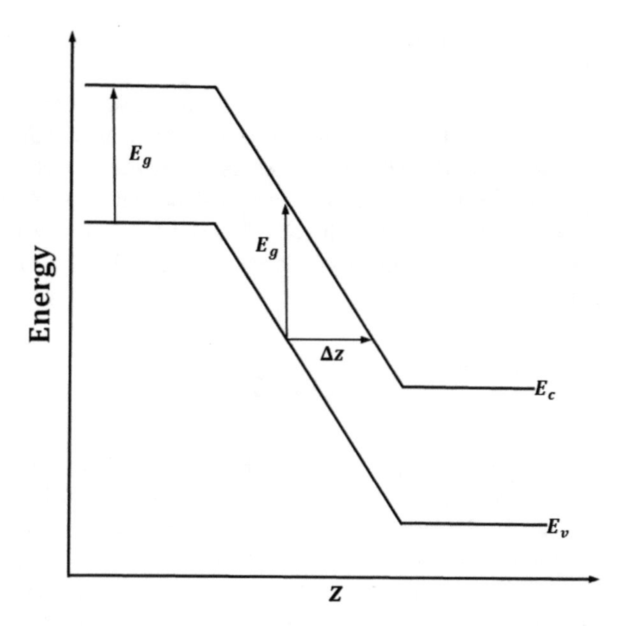

Fig. 3.2 Illustration of a triangular barrier. Δz is the difference between the turning points, corresponding to the length that the electron has to travel to tunnel through the triangular barrier

the trap potential. The mass matrix has been approximated to be [13] $M^2 \left(\frac{m_{eff}}{m_0}\right) \sim 10^{-23}$ eV^2 cm^3, where m_0 is the mass of a free electron.

Both tunneling currents increase with increasing cutoff wavelength, and/or increasing bias. Figure 3.2 shows an illustration of the triangular barrier that the electrons would have to tunnel to conduct. The distance Δz that the electrons have to tunnel through for this triangular barrier is [10] $\Delta z = E_g/qE$, where E_g is the band gap energy, q is the charge of an electron, and E is the electric field $\left(\propto \sqrt{E_g - qV_{detector\ bias}}\right)$. For an array of a given cutoff wavelength, increasing the applied reverse bias ($V_{detector\ bias} < 0$) will reduce the distance that electrons need to tunnel, therefore increasing the probability that electrons will tunnel either from the valence band, or from a trap state to the conduction band.

Data from pixels that appear to exhibit trap-to-band tunneling in the LW10 [17, 20, 21] and the LW13 devices have also shown a soft breakdown in the I-V curves, prior to the onset of trap-to-band tunneling. Other authors have observed soft breakdowns in 4H-SiC [36, 37] and Silicon [38] diodes caused by screw dislocations and stacking faults respectively, which became electrically active at a certain "threshold voltage". Furthermore, Neudeck et al. [36], Ravi et al. [38], and Benson et al. [39] (for HgCdTe on Si) showed that there is no correlation between the degraded I-V characteristics and the trap density, because all traps associated with dislocations may not contribute to the soft breakdown due to the varying threshold voltage for dislocations within the same diode at which they become electrically active.

Though we cannot say with certainty what type of dislocation [27] causes the soft breakdown in the LW13 arrays presented here, we believe a defect associated mechanism, likely different from that described by the authors above is responsible for this early soft breakdown and the onset of trap-to-band tunneling as traps become electrically active. Bacon [20] parametrically fit the I-V curves by introducing a threshold voltage at which certain traps become active and contribute to trap-to-band tunneling as

$$n_t = n_{t_i} + \frac{n_{t_d}}{1 + exp\left[\frac{\gamma q(V_a + V_{detector\ bias})}{kT}\right]}, \tag{3.26}$$

where n_{t_i} is an initial active trap density, n_{t_d} is the trap density due to activated dislocations at a voltage V_a, and γ is a parameter which dictates how sharply the current increases due to the soft breakdown before reaching the current expected from trap-to-band tunneling. A small modification to the Eq. 2.29 in Bacon [20] was made where we multiply the sum of the activation voltage and the detector bias by γ (Bacon multiplied $V_{detector\ bias}$ by the scaling factor γ). This change was made so that the fitted γ parameter only affected the sharpness of the soft breakdown, and not the fitted activation voltage.

Although this is not a physical model since we do not know the trap identity, a more detailed study using this phenomenological fit can give us information about the traps such as the trap density and a distribution of trap energies.

Chapter 4
Array Characterization

To characterize the performance of the LW13 and LW15 arrays, the first step is to calibrate the data. Calibration involves the calculation of the multiplexer gain, diode capacitance to determine the ADU to electron conversion, and non-linearity measurements to account for the changing capacitance in the diode with integrated signal. For the LW13 arrays, all of the calibration measurements were performed at a temperature of 30 K, and at 25 K for the LW15 arrays.

The performance of the arrays is characterized by the dark current, well depth, and read noise. Dark current and well depth measurements were obtained per pixel at stable temperatures ranging from 23 to 36 K with applied reverse bias ranging from 50 to 400 mV. In addition to the stable temperature measurements, dark current with slowly increasing temperature was measured to fit dark current models to data as a function of temperature. Read noise was measured per pixel for two LW13 devices, and for all three LW15 devices.

For the first phase of this project, the main focus was to determine the diode structure that would be best suitable to increase the cutoff wavelength for the second phase of this project. Since dark current played a major role in limiting the performance of the LW13 arrays, our imposed operability requirements revolved around dark current and well depth.

Operability maps are used as a way of identifying any spatial dependence on poor performing pixels. Often, if an entire row or column is inoperable, those inoperabilities are most likely due to an issue with the multiplexer, and not the photosensitive material. Bonding processes may also affect the operability of devices; such as cases when too much pressure is applied in the bonding of the photosensitive material to the multiplexer, causing excess dislocations due to external applied forces. An intrinsic occurring cross-hatching pattern in all HgCdTe devices grown on CdZnTe substrates can also be identified using the operability map. All devices presented here exhibited a cross-hatching pattern as inoperable pixels (large dark current and/or low well depth). This pattern arises from the crystal axis mismatch between HgCdTe and the substrate, and will be further discussed in

M. Cabrera, *Development of 15 Micron Cutoff Wavelength HgCdTe Detector Arrays for Astronomy*, Springer Theses, https://doi.org/10.1007/978-3-030-54241-2_4

Sect. 4.5. In order to correctly identify and match the cross-hatching to the known directions of this pattern, the relative angles between the cross-hatching lines were measured using the Hough transform.

Data acquisition and reduction methods for each of the data sets are described in the following sections.

4.1 Reference Pixel Subtraction

The reference pixels in the perimeter of the H1RG mux are used to correct for any bias drifts or row noise in the pixel circuitry. As it was mentioned in Sect. 1.4, the light sensitive diode material is replaced by a capacitor in the reference pixels; therefore, any behavior as a function of time seen in the reference pixels will be shared by the light sensitive pixels since they share the same underlying circuitry, and can be corrected for using the reference pixels.

To correct the data using the reference pixels, the eight reference pixel columns (four right-most and left-most on the array) are averaged in the column direction, resulting in a single column with 1016 rows (the top and bottom four rows of reference pixels are discarded). This averaged reference pixel column is then convolved with a smoothing filter[40] and subtracted from the rest of the columns with photon-sensitive material to correct for any drifts, and will also reduce the total noise in each corrected frame.

All data presented here has been reference pixel subtracted, except for the capacitance data for which the noise in each frame is of interest.

4.2 Calibration

4.2.1 Multiplexer Gain

The voltage at the integrating node due to the collected charges in the p-n junction of photodiodes is the input referred signal. When the voltage is read out, there are two FETs in the path of the readout in the multiplexer (see Sect. 2.3) that amplify that signal (subsequently, the signal is further amplified by the external array controller, see Sect. 2.2). The output voltage that is read out, is not the actual voltage (or signal) at the integrating node, and is referred to as the output referred signal.

Converting between the output and input referred voltage requires the knowledge of the multiplexer gain. To measure the multiplexer gain G_{mux}, the voltage at the integrating node must be known, and read out. The reset voltage is used as the input voltage to measure the multiplexer gain. To prevent the pedestal injection from changing the known input voltage from the reset, the reset switch is kept closed while reading it out. Also, to prevent the diode from debiasing and changing the

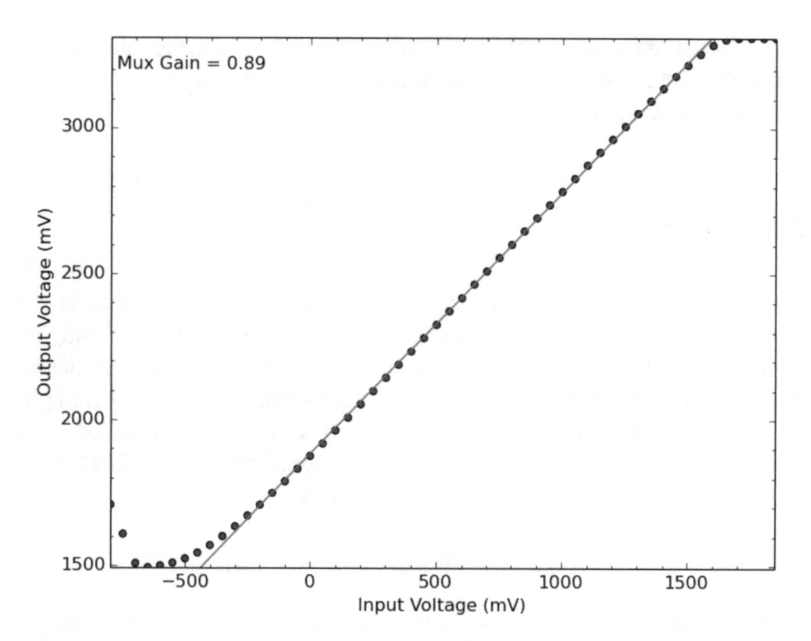

Fig. 4.1 Output *vs.* input referred signal curve used to calculate the gain from the multiplexer. The input referred signal corresponds to the applied reset signal, where the reset switch remained in the closed position while reading out the voltage at the integrating node. The multiplexer gain corresponds to the slope in the linear region of the curve. The devices presented here will operate between 100 mV (reset voltage) to the bias applied at the substrate (typically can be as large as 450 mV, corresponding to 350 mV of applied reverse bias)

voltage at the integrating node, the data are obtained with the cold dark shutter in place, and the applied voltage on the substrate is equal to the reset voltage, making the applied voltage zero to eliminate dark currents that can debias the diode.

The ratio of the output to the input signal is the multiplexer gain. The input referred signal is then given by

$$S_{input} = \frac{S_{output}}{G_{mux} G_{external}}, \tag{4.1}$$

where $G_{external}$ is the gain from the external array controller. Figure 4.1 shows the output *vs.* input referred signal curve used to calculate the multiplexer gain. In addition to calculating the multiplexer gain, this data set is also used to ensure that data is taken in the linear regime of the FETs.

To test the arrays presented here, a reset voltage of 100 mV is always applied, with the substrate voltage dictating the applied bias across the detector. As it was mentioned in Sect. 2.3, for reverse bias operation, the voltage at the substrate must be larger than the voltage at the integrating node, opposing the built-in voltage polarity. The voltage at the integrating node will therefore vary between the 100 mV applied to reset the array, and the voltage at the substrate. Typically, the largest applied bias that these arrays are tested with is 350 mV, making the substrate voltage 450 mV. The data shown in Fig. 4.1 show that the array presented would be operated within the linear regime of the FETs.

4.2.2 Capacitance

The first step in calibrating the data is to measure the nodal capacitance to convert the signal from ADUs to electrons. The nodal capacitance is measured using the σ^2 vs. signal method[33]. This method takes advantage of the Poisson statistics nature of photons, where the noise in the incident photons (shot noise) for a given pixel is $\sqrt{S_{[e]}}$, where S is the number of collected charges by a specific photodiode, and the subscript in brackets will denote the units of the specified quantity. The total noise for a given pixel adds in quadrature, and is given by

$$\sigma^2_{Total,\ [e]} = \sigma^2_{System,\ [e]} + \sigma^2_{Shot,\ [e]}, \tag{4.2}$$

where the system noise includes any noise related to the electronics, the multiplexer, and the dark current in the diode. If the array is operated in a regime in which the dominant component in the noise is the shot noise (i.e. photodiode signal is photon flux limited), using Poisson statistics, the total noise squared (i.e. variance) is

$$\sigma^2_{Total,\ [e]} = S_{[e]}. \tag{4.3}$$

The output signal is not the number of charges collected in the p-type implant, but the voltage at the integrating node due to the collected charges. To convert the signal and noise between electrons and volts across a single photodiode, the following relations are used:

$$S_{[V]} = \frac{S_{[e]}q}{C} \tag{4.4}$$

$$\sigma_{[V]} = \frac{\sigma_{[e]}q}{C}, \tag{4.5}$$

where q is the elementary charge, and C is the pixel capacitance. Combining the above relations with Eq. 4.3, the pixel capacitance is given by

$$C = \frac{q S_{[V]}}{\sigma^2_{Total,\ [V]}}. \tag{4.6}$$

4.2.2.1 Capacitance Data Acquisition

To measure the capacitance, photon signal at various fluence levels is needed, where the noise can be measured by calculating the rms noise in many measurements taken for each of the fluence levels.

There are two ways to measure different fluences. One is to expose the array to light passing through any of the filters, and increasing the integration time. Since we want to cover a large range of fluence levels, and taking multiple frames to calculate rms noise, the total data acquisition time for this set will depend on the number of fluence levels that are used. The total time can become quite long using this method, and thus 1/f noise may become a contributing factor skewing the results.

The second method to accomplish this, and the method that is used here, is to use the circular variable filters in our dewar setup since the room temperature blackbody radiation flux that the array is exposed to varies with wavelength. Therefore, the integration time is kept the same, and only the position of the filter is changed to obtain different fluence levels. Sets of 100 CDS images, each set at varying fluence levels, were used to obtain the signal and the rms noise.

4.2.2.2 Capacitance Calculation

Each set of 100 frames corresponding to the same wavelength/fluence level is averaged to obtain the signal, and the rms noise is calculated from the 100 frames. The raw noise squared *vs.* average signal for a pixel in LW13 array H1RG-18508 is shown in Fig. 4.2. The fitted slope gives the ratio of the noise squared and the signal, where the inverse of the slope is used in Eq. 4.6 to calculate the capacitance $\left(slope^{-1} = S_{output,\ [ADU]}/\sigma^2_{output,\ [ADU]}\right)$. Since the raw data shown on Fig. 4.2 correspond to the output referred signal in units of ADUs, the signal and noise squared must be converted to the input referred signal in units of volts (see Sect. 2.2). The capacitance is calculated for each pixel in the array using the expression

$$C = \frac{q S_{output,\ [ADU]}}{\sigma^2_{output,\ [ADU]}} G_{mux} G_{external} \frac{2^{16} ADU}{5V}. \tag{4.7}$$

This capacitance is not corrected for interpixel capacitance and can therefore overestimate the electron per ADU conversion.

4.2.3 Interpixel Capacitance

Interpixel capacitance (IPC) is an effect in which signal from a pixel is capacitively coupled to neighboring pixels, causing an apparent signal exchange between pixels. This apparent exchange of signal is a change on the electrostatic potential between

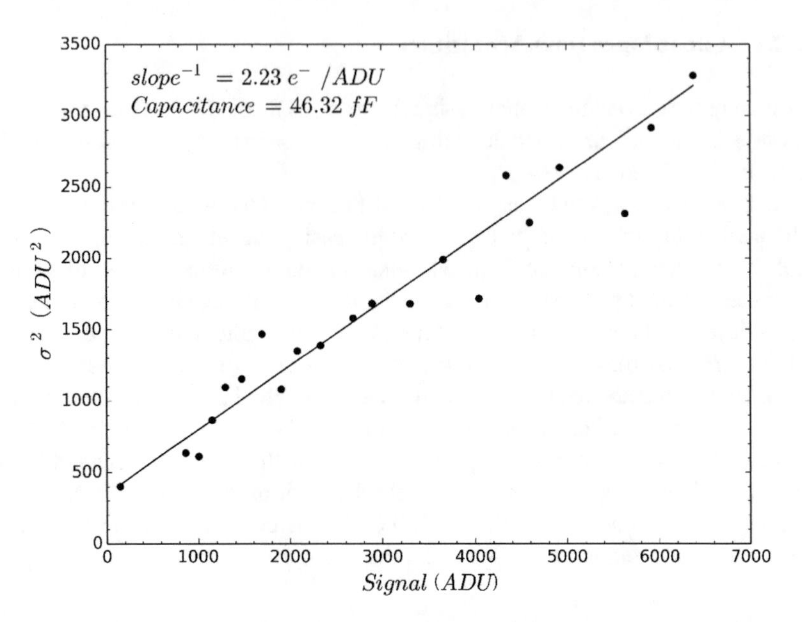

Fig. 4.2 Noise squared *vs.* signal plot for a pixel in H1RG-18508 at a temperature of 30 K and 150 mV of applied bias, where the slope of the fitted line corresponds to the conversion factor between ADUs and e^-. The capacitance calculated here is not yet corrected for interpixel capacitance. At a signal of zero, the y-intercept is equal to the read noise for this pixel

neighboring pixels at their respective integrating nodes. Since the noise associated with the collected signal is due to shot noise from the incoming photons, any added signal to a pixel due to IPC will be noiseless. This underestimation of the noise leads to an overestimation of the nodal capacitance[41], resulting in an overestimation in the conversion gain of electrons to ADU.

The nature of the IPC can be described as a convolution of the image with a blur kernel. For a given pixel, if it is assumed that IPC only affects its nearest neighbors, the blur kernel can be described by

$$K_{i,j} = \begin{pmatrix} 0 & \alpha & 0 \\ \alpha & 1 - 4\alpha & \alpha \\ 0 & \alpha & 0 \end{pmatrix}, \tag{4.8}$$

where α is coupling parameter between a given pixel and its four nearest neighbors. This assumes an equal amount α of signal of a given pixel is distributed among its four nearest neighbors, ignoring the diagonal nearest neighbors, leaving the central pixel with a signal fraction of $1 - 4\alpha$.

Moore et al. [41] showed that the IPC can affect the noise squared by as much as 10%, and used the autocorrelation method to show that the noise squared corrected for IPC is

$$\sigma^2_{IPC\ corrected,\ [ADU]} = \frac{\sigma^2_{Output\ measured,\ [ADU]}}{(1-8\alpha)}. \tag{4.9}$$

The above expression is valid for small coupling parameter α. The nodal capacitance is corrected for IPC by substituting the IPC corrected noise squared into Eq. 4.7.

Donlan et al. [42] have shown that IPC is a function of the signal for a central pixel, and that of the four nearest neighbors. To measure the coupling parameter as a function of signal, single pixel reset is required to isolate specific pixels. In this work, an approximation of a constant coupling parameter is applied to correct for IPC. Our group is currently working to characterize IPC as a function of signal.

4.2.3.1 IPC Coupling Parameter Calculation

The nearest neighbor method was used to determine the IPC coupling parameter; Pixels with very high dark current (hot pixel) were used to determine the coupling parameter α.

Considering a central hot pixel and its surrounding nearest neighbors, the output signal of the center pixel is given by

$$S_{Center\ Output} = S_{Center\ Actual} - 4\alpha S_{Center\ Actual} + 4\alpha S_{Background}, \tag{4.10}$$

where $S_{Center\ Actual}$ is the actual signal charge collected in the central pixel, and $S_{Background}$ is the background signal. In an ideal array without IPC effects, the dark current on a hot pixel should not affect neighboring pixels; therefore, the neighboring pixels should have a signal equal to the background. The central pixel will appear to have lost $\alpha S_{Center\ Actual}$ to each of the four nearest neighbors due to IPC, and gained $\alpha S_{Background}$ from each of the nearest neighbors. By the same reasoning, each of the four nearest neighbors will have an output signal of

$$S_{Neighbor\ Output} = S_{Background} - \alpha S_{Background} + \alpha S_{Center\ Actual}. \tag{4.11}$$

Solving the set of equations for the coupling parameter, α is given by

$$\alpha = \frac{S_{Neighbor\ Output} - S_{Background}}{S_{Center\ Output} + 4S_{Neighbor\ Output} - 5S_{Background}}. \tag{4.12}$$

Using the lowest fluence level set of 100 CDS images taken to compute the capacitance, and with reference pixel subtraction, an α value is calculated for all hot pixels (signal level six standard deviations above the mean) in the 100 images. In this data set, each of the neighboring pixels for any given hot pixel may not have the same signal; therefore, the mean coupling parameter is calculated, and given by

$$\langle \alpha \rangle = \frac{\langle S_{Neighbors\ Output} \rangle - S_{Median\ Background}}{\left(S_{Center\ Output} - S_{Median\ Background}\right) + 4\left(\langle S_{Neighbors\ Output} \rangle - S_{Median\ Background}\right)}, \tag{4.13}$$

where $S_{Median\ Background}$ is the local median signal value of the perimeter 5 × 5 pixels (assumed to not be affected by IPC from the central hot pixel), and $\langle S_{Neighbors} \rangle$ is the mean output signal from the four nearest neighbors. From the coupling parameter distribution in the array, the median value is used to correct the nodal capacitance, where Eq. 4.7 is multiplied by the correction factor of $1 - 8\,\langle \alpha \rangle$.

4.2.4 Non-linearity

Using the nodal capacitance to convert the signal at the integrating node to electrons is only valid for low signal data. For large signals, the variation of the diode capacitance needs to be accounted for. As the diode debiases with the collection of signal, the reduction of the junction depletion region increases the nodal capacitance[43]. Using a smaller capacitance to convert the signal to the number of collected charges can lead to an underestimation of the actual number of charges in the p-type implant; therefore, a correction of the nodal capacitance as a function of collected signal is needed to accurately calibrate large signal images. Although the main focus of this work has been to characterize the dark current and well depth of these devices, and therefore the non-linearity is not used here to correct for any large signal images, it is important to show any potential difficulties in correcting for the non-linear capacitance.

To measure the non-linearity of these devices, the signal collected from a constant illuminating flux is sampled in SUTR mode until saturation is reached. Characterizing the non-linear capacitance at small and large signals is important to identify any potential sources that can complicate the calibration of the non-linear capacitance such as dark current at low signals, or the "brighter-fatter" effect[44] at large signals. The 3.3 μm filter is placed in the illumination path since it is one of the narrowband filters with the lowest emerging signal flux from the room temperature black-body source used to illuminate the array.

Non-linearity sets with different applied biases are taken consecutively, where for convenience, 270 SUTR frames with an integration time of 5.8 s are taken for the different applied biases to ensure the tested array reaches saturation regardless of the well depth.

Each of the 270 frames in a non-linearity data set are initially reference pixel subtracted. The non-linear collected signal rate is then calculated by dividing the signal over the time it took to collect the signal, and normalized to the first sampled signal. The signal is not corrected for dark current.

The normalized signal rate is also referred to as C_0/C, where C_0 is the nodal capacitance at zero collected signal and C is the nodal capacitance. To correct for the non-linear capacitance, the slope of a fitted line to the non-linearity curve (between

20 and 80% of the saturating signal) is used. The difference between unity and the non-linearity curve corresponds to the fraction of signal that is not accounted for by assuming a constant capacitance.

4.3 CDS Read Noise

To compute the read noise, 64 SUTR sets (each SUTR consisting of 36 frames) were taken with the cold dark shutter in place with an integration time of 5.5 s between each sample. From this, 64 Fowler-1 (CDS) images were created, where the total rms noise was calculated. The total CDS noise when measured in the dark is expected to be dominated by the read noise.

Although only the first two frames of each SUTR set are used to form the CDS images from which the rms noise is computed, the rest of the frames are used to estimate the dark current, and identify any anomalies present by comparing this measured dark current with the that calculated from the data set taken to characterize the dark current and well depth (see the following section).

4.4 Dark Current and Well Depth

Dark current and well depth have become an integral part of the characterization of these arrays since large values can set a limit on the observations that can be made with these devices. Measurements of dark current and well depth are measured together here since both quantities are coupled. Pixels with large dark currents can considerably deplete the well depth between reset and the pedestal frame, and in cases of very large dark currents, the pixel may be saturated before the pedestal frame is read out. Furthermore, as it is shown in Chap. 3, the initial dark current can be a strong function of the initial detector bias (well depth): Pixels that have almost completely debiased before the pedestal frame will appear to have low dark currents. If only dark current measurements are considered to determine operability, there is a degeneracy among low dark current pixels: The well depth is used to break this degeneracy.

Moreover, to correctly characterize pixels with large dark currents that have a strong bias dependence, dark current and well depth must be measured sharing the same reset. The well depth (actual initial bias) can vary depending on the dark current and the pedestal injection (see Sect. 2.3). The pedestal injection can vary from reset to reset, and this variation in the bias can affect the initial dark current in pixels that exhibit tunneling current. Measuring the dark current and well depth under the same reset removes any ambiguity of the initial applied bias.

4.4.1 Dark Current and Well Depth Data Acquisition

To measure the dark current, 200 full frames are read out in SUTR mode in the dark with an integration time of 5.8 s between samples. The initial dark current is given by the slope at the beginning of the dark SUTR curve.

To determine the well depth, as was mentioned in Sect. 2.3, the zero-bias point with no illumination (V_{ZBP}) must be measured. The well depth is given by the difference between the zero-bias point and the pedestal, corresponding to the initial detector bias. Since the dark current can often have very small values ($<1\ e^{-}/s$), it is impractical to allow the array to reach saturation (zero-bias point) with dark current alone.[1] Following the 200 samples in the dark, without resetting the array, the filter wheel is moved from the cold dark shutter to the 8.6 μm filter to allow a radiative flux to saturate the array. 50 samples under illumination are taken with an integration time of 25 s between samples, where the saturating level (V_{OC}) reached due to the signal flux includes a contribution of ~10–20 mV of forward bias[20]. To allow the bias across the diode to settle back to the zero-bias point from forward bias, the cold dark shutter is placed back into the illumination path to place the detector back in the dark. Again, without resetting, 200 additional samples are taken in the dark with an integration time of 5.8 s.

Figure 4.3 shows the SUTR curve of a single pixel used to measure the dark current and the well depth. The very first sample in the ramp is the pedestal, which can be used to normalize the entire SUTR curve to a zero signal point.

4.4.2 Dark Current Calculation

The dominant component of dark current can change the behavior of dark current of a pixel as it debiases, namely constant or variable behavior as a function of bias. An algorithm was developed to identify the linear *vs.* non-linear dark charge behavior when calculating the dark current. In constant dark current cases, a line is fitted to the dark SUTR data, where the slope of the line is the dark current. The 200 data points taken to measure the dark current are not always used to fit the slope; The algorithm uses the double derivative of the signal *vs.* time SUTR curve to identify cosmic ray events, and cases when burst noise[20] is present. A cosmic ray hit or burst noise will cause a discontinuous increase in the collected signal *vs.* time, showing up as a large deviations from a value of zero in the double derivative. If either event occurs during a linear part of the collected dark charge *vs.* time ramp, the slope before and after the event should be very similar. For simplicity, only the portion of the SUTR curve before those events is used to calculate the slope.

[1]For the best performing LW13 array, it could take up to ~60 h to reach saturation if only charges from dark currents are collected.

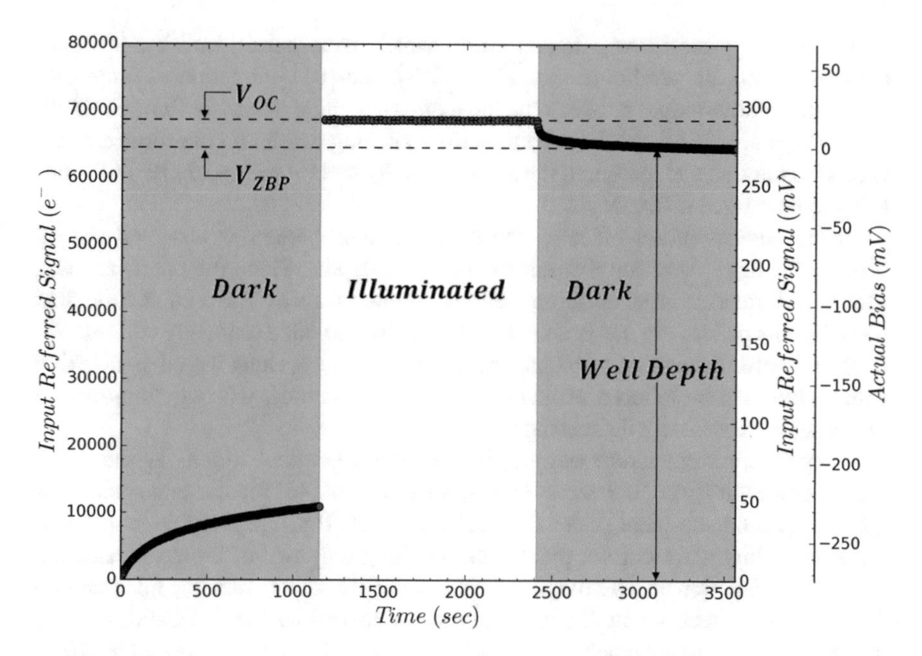

Fig. 4.3 Signal *vs.* time taken to measure dark current and well depth. The three different sections of the SUTR curve are all taken under the same reset

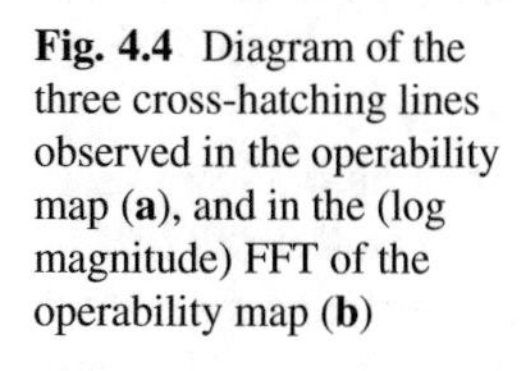

Fig. 4.4 Diagram of the three cross-hatching lines observed in the operability map (**a**), and in the (log magnitude) FFT of the operability map (**b**)

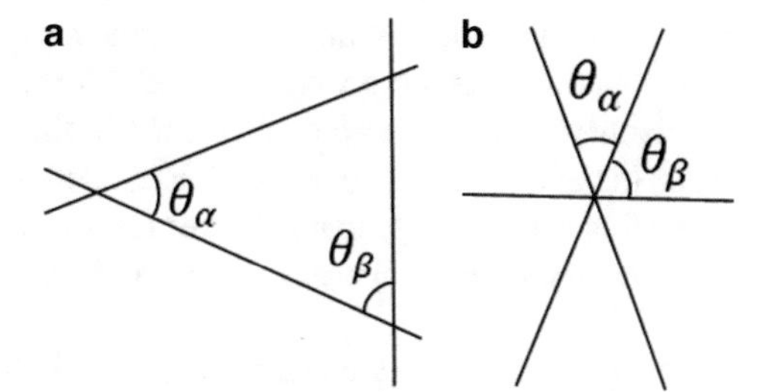

For non-linear cases, due to the changing curvature of the SUTR in the dark as the diode debiases, the initial dark current is approximated by the difference between the first two dark signal points, divided by the integration time.

4.5 Cross-Hatching

The lattice mismatch between the CdZnTe substrate and the molecular-beam epitaxy (MBE) grown HgCdTe detector bulk material causes the formation of misfit dislocations along three particular directions, forming a cross-hatching pattern. The misfit dislocations form along the intersection of the {111} slip planes in zinc blende crystals and the growth plane of these arrays[29, 30, 45].

The set of three cross-hatching lines, which have been identified by other authors[29, 30], lie parallel to the $[\bar{2}31]$, $[\bar{2}13]$, and $[01\bar{1}]$ directions. Figure 4.4a shows the relative angles between the three cross-hatching lines. Martinka et al.[29] and Chang et al.[30] have observed the cross-hatching pattern on the surface of the HgCdTe material, and measured the angles to be $\theta_\alpha \approx 44-45°$ and $\theta_\beta \approx 67.5-68°$ using atomic force microscopy.

In the devices presented here, the cross-hatching pattern is observed among pixels with large dark currents and/or low well depths. Since the cross-hatching pattern is formed from dislocations, all inoperable pixels at low temperatures that lie along this pattern are inoperable due to large trap-to-band tunneling currents. G-R currents also depend on trap characteristics, but do not show the cross-hatching pattern, and can be reduced at low operating temperatures, whereas trap-to-band tunneling increases with decreasing temperature.

Large dark current and/or well depth is not the only effect seen along the cross-hatching lines in HgCdTe detector arrays. Shapiro et al.[46] see the same pattern as QE variation on sub-pixel scales on a 2.5 μm cutoff HgCdTe array, in addition to a cluster of high dark current pixels that lie along only one of the cross-hatching directions. Variation in QE measurements along the cross-hatching lines can be due to trapped electrons in empty energy states created by the dislocations. If the trapped electrons do not reach the p-n junction, this will appear as a lower QE effect. In addition to affecting QE measurements and dark currents, traps may also be responsible for latent images, where a fraction of the signal from a source may still appear in images taken after resetting the array. Latent signal comes from released electrons that filled empty energy levels from traps or dislocations.

Identifying the cross-hatching lines in the devices presented here allows us to make the assertion that pixels with large dark current and/or low well depth that lie along this pattern are probably inoperable due to the presence of dislocations created from the crystal axis mismatch between the CdZnTe substrate and the HgCdTe bulk material.

Since not all pixels along a certain cross-hatching line are inoperable, in addition to also having randomly distributed inoperable pixels not associated with the cross-hatching, it can be difficult to identify specific lines in the operability map. Taking the two dimensional Fast Fourier Transform (FFT) of the operability map shows the cross-hatching pattern clearly (only rotated by 90°). Figure 4.7 shows an example of a simulated map of pixels distributed along cross-hatching line directions and randomly distributed pixels along with the FFT of the simulated map. The unmistakable cross-hatching pattern in the FFT image is used as the input to the Hough transform to determine the angles between the cross-hatching lines, and matching those directions to the known cross-hatching patterns in HgCdTe.

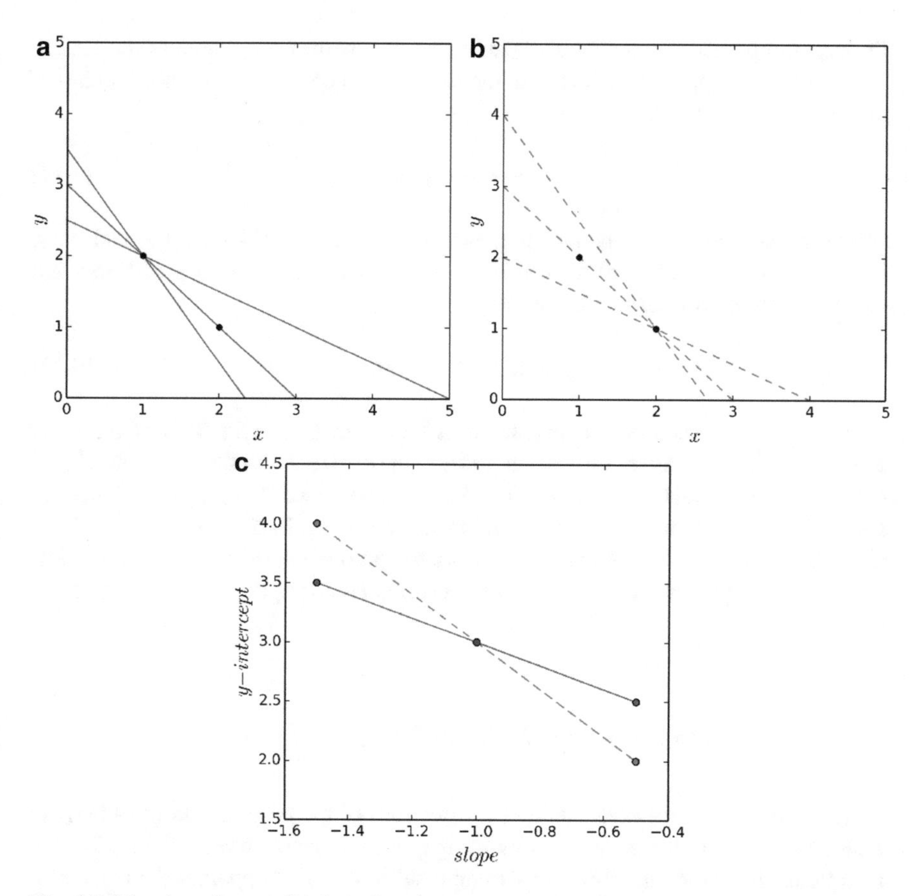

Fig. 4.5 Mapping of two points into slope-intercept space. (**a**) shows three of the many possible lines that pass through point (1,2), while (**b**) shows three lines that pass through point (2,1). Each of those lines is mapped to a single point in slope-intercept space (**c**). Each data point in x-y space corresponds to a line in slope-intercept space, where the intersection of the two lines, corresponds to the slope and y-intercept of a line that passes through both points

4.5.1 Hough Transform

The Hough Transform is commonly used to identify lines in images by mapping each individual pixel of interest (inoperable pixels in our case) into slope-intercept parameter space. Figure 4.5 shows a simple case where two points with Cartesian coordinates get mapped into two lines in slope-intercept space. The intersection of the two lines corresponds to the slope and y-intercept of a line that passes through both points. If additional co-linear points are added, more lines in parameter space would intersect at the same point in Fig. 4.5c.

The Hough Transform is a two-dimensional histogram in parameter space, where each bin is equal to the counts of slope-intercept pairs for lines that pass through the points of interest. The parameter pair with most counts in the Hough Transform corresponds to a line with the most co-linear points. The shortcoming of parameterizing a line as

$$y = mx + b, \tag{4.14}$$

where m and b are the slope and y-intercept respectively, is that for vertical lines, the slope diverges. A line of the form of Eq. 4.14 can be re-parameterized (to avoid slopes of infinity for vertical lines), as

$$\rho = x\cos\theta + y\sin\theta, \tag{4.15}$$

where ρ is the perpendicular distance to a line from the origin (0,0) of the input image, and θ is the anti-clockwise angle from the horizontal to the perpendicular of the same line.[2] Hough space refers to the re-parameterized $\theta - \rho$ space. Figure 4.6 shows the same two points from Fig. 4.5 in Hough space. Each point is mapped to a sinusoid in Hough space, allowing for characterization of vertical lines. All possible lines that can pass through a point can be described with angles to the perpendicular ranging from $-90° < \theta \leq 90°$.

4.5.2 *Cross-Hatching Angle Calculation*

To show the method of determining the directions of the cross-hatching pattern, an image is created to mimic an operability map with inoperable pixels along known directions. Figure 4.7a shows an example with 4% of "inoperable" pixels in a 1016×1016 pixel image distributed along lines with an angle of $\theta_\alpha = 45°$ between the cross-hatching directions $\left[\bar{2}31\right]$ and $\left[\bar{2}13\right]$, while 1% of pixels are distributed along the $\left[01\bar{1}\right]$ cross-hatching (shown vertically here, $\theta_\beta = 67.5°$). Three percent of "inoperable" pixels have also been randomly dispersed on the entire test image. To properly create the cross-hatching pattern, 50 lines with the desired directions were chosen a priori, and the percentage of pixels to be distributed along these lines are randomly dispersed along those lines.

Figure 4.7b shows the FFT of the simulated operability map. The cross-hatching lines on the FFT image are rotated by 90°, and are shown by a distinct set of lines. Applying the Hough transform to the FFT image will reduce the number of falsely identified lines. To use the FFT image as the input to the Hough transform, it must be

[2] Since all points of interest lie on the first quadrant, i.e. positive x and y coordinates, lines with a positive slope passing through those points will have its perpendicular at an angle > 90°. Since the Hough transform here is chosen to be limited in the –90° to 90° range, lines with a perpendicular lying in quandrant II, are instead mapped to a parallel line in quadrant IV. This parallel line would then have a negative distance ρ to the actual line passing through the points of interest.

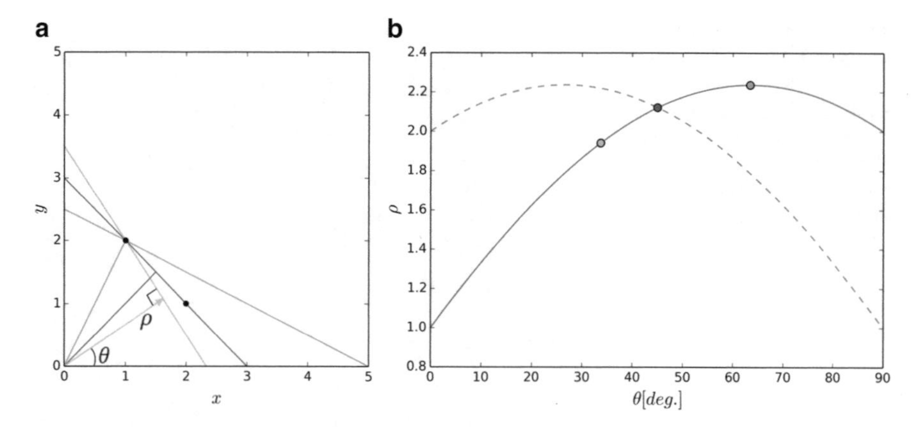

Fig. 4.6 Mapping of two points into Hough space ($\theta - \rho$). (**a**) shows the same three lines from Fig. 4.5a that pass through point (1, 2). Each of those lines is mapped to a single point in Hough ($\theta - \rho$) space (**b**), where θ and ρ are the angle and length of the perpendicular to the line to be parameterized. Each data point in x-y space is mapped as a sinusoid in Hough space. In (**b**), the solid and dashed line are the same lines in Fig. 4.5c, where the intersection of the two lines corresponds to the parameters of the perpendicular to the line that passes through both points in (**a**)

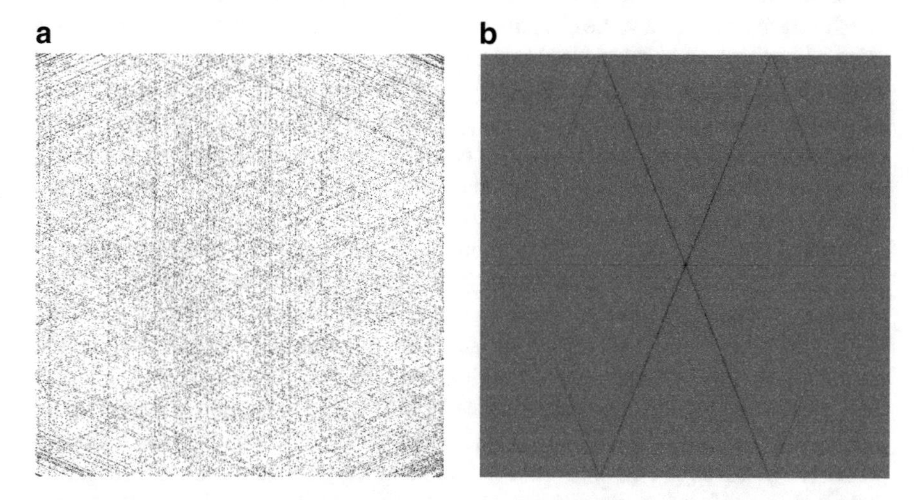

Fig. 4.7 Simulated inoperability map and its FFT. Four percent of pixels in a 1016 × 1016 image, shown as black in (**a**), have been distributed along the two cross-hatching lines shown in Fig. 4.4 with $\theta_\alpha = 45°$ and 1% of pixels along the vertical directions. An additional 3% of pixels were randomly distributed on the entire image to mimic an operability map of our devices, only with known directions of the cross-hatching lines. (**b**) shows the log magnitude FFT of (**a**), where the cross-hatching lines are clearly shown, only rotated by 90°. (**b**) has been scaled to show a better contrast between the background and the cross-hatching features

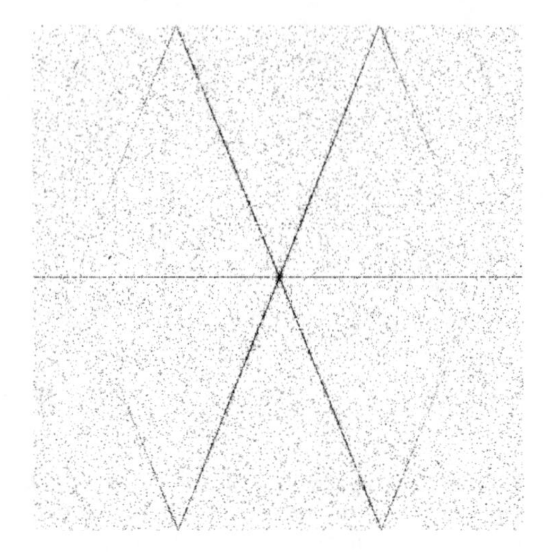

Fig. 4.8 Binarized input image to Hough transform. A threshold of 1.5 standard deviations larger than the mean was used on Fig. 4.7b. Pixels above the threshold were assigned a value of zero, and a value of unity to the rest

binarized. The cross-hatching pattern in the FFT image has a higher value than the background, and a threshold of 1.5 standard deviations above the mean is chosen to isolate the cross-hatching pattern from the background. The pixels above the threshold are assigned a value of zero and unity for the rest. Only those pixels with a value of zero will be mapped into Hough space. Figure 4.8 shows the binarized input image to the Hough transform. It is important to note that the output angles from the Hough transform (with the FFT image as the input) coincide with the angles (w.r.t the horizontal) of the simulated cross-hatching pattern. Therefore, from here on, the output angles from the Hough transform will be denoted as the angles from the cross-hatching lines.

From Fig. 4.9, it can be seen that there is a large number of sinusoids crossing at angles $\sim -22.5°$ and $\sim 22.5°$. Though not clear from the figure, there is also a high concentration of sinusoids crossings near $\pm 90°$. Multiple parallel lines in the FFT image will appear as a cluster of sinusoid crossings at similar angles, only at different values of ρ. Since only the angle of the FFT lines/cross-hatching pattern is of interest, for each binarized point in the FFT image, a histogram is formed from the angle of the perpendicular to the line that each of those points is most likely a part of (largest Hough transform count). To reduce the noise in the angle distribution that would be added from the points that are not part of a line feature, only the angles that have a count five standard deviations above the mean in the Hough transform are used. This threshold will eliminate the contribution from the large concentration of pixels (from the intersection of several line features or circular features) in the center that will be co-linear with randomly distributed points, increasing the Hough transform value for those angles above the five standard deviation from the mean threshold.

Figure 4.10a shows the perpendicular angles (matching the cross-hatching direction) distribution that each of the binarized pixels belongs to, according to the Hough transform. From this distribution, three clear set of lines can be identified.

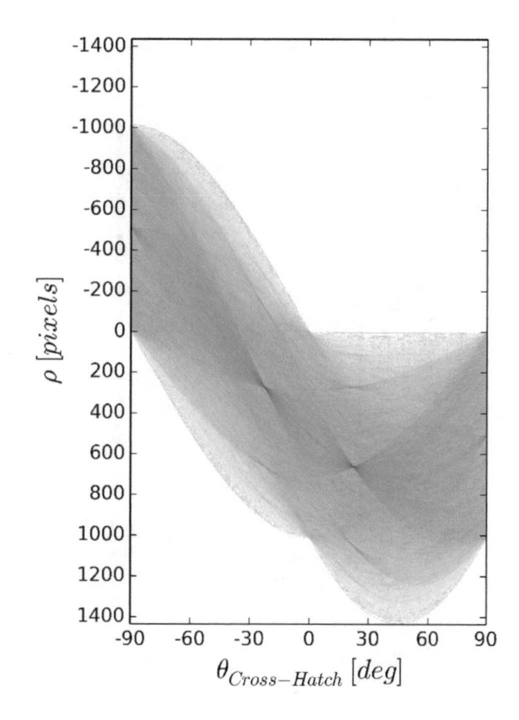

Fig. 4.9 Hough transform of Fig. 4.8. The input image to the Hough transform is a binarized FFT image, and because the FFT image shows the cross-hatching pattern rotated by 90°, the output angles of the Hough transform correspond to the actual cross-hatching pattern w.r.t. the horizontal

The median of the three perpendicular angle distributions are −22.5°, 22.5°, and 90°. The distribution of lines near ±90° correspond to the horizontal line in the FFT. To get the median of this distribution, the perpendicular angle distribution is converted to the angles of the actual lines in the FFT (shown in Fig. 4.10b). This places the distribution of angles near 0°, where the median of the distribution can now be calculated.

The median angles of the three different distributions in Fig. 4.10a are the angles that were used to create the simulated operability map. The median of the angle distributions in Fig. 4.10a and b are overlaid on the simulated operability map and its FFT in Fig. 4.11a and b respectively to show the correct identification of the pattern direction.

4.6 Dark Current *vs.* Bias

Dark current *vs.* bias (I-V) curves are used to fit dark current models to data. The data set taken to determine the initial dark current and well depth also allows us to determine the dark current and detector bias per pixel at any time in the dark integration ramp as pixels debias. The I-V curve is given by the slope of the dark current SUTR data versus the detector bias voltage (e.g. Fig. 4.3). In cases when the dark current is small and linear, the dark current SUTR curve will span less than

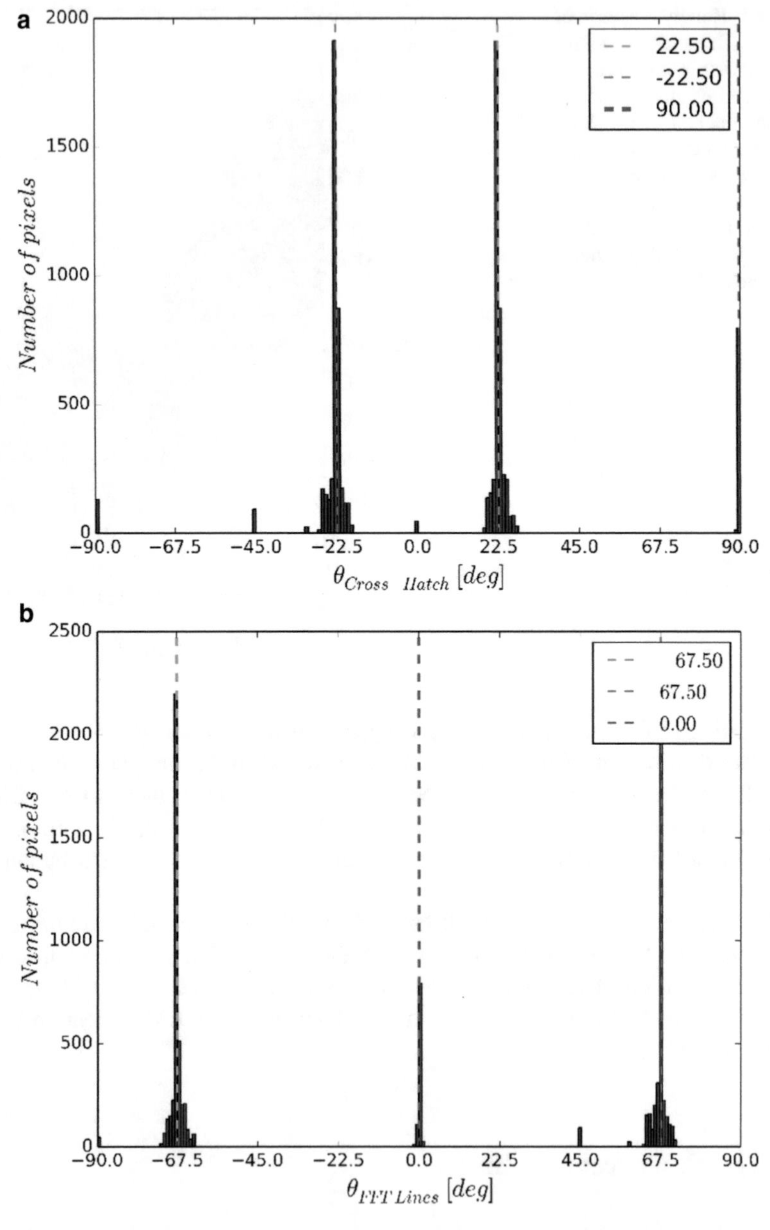

Fig. 4.10 One-dimensional angle histograms of possible cross-hatching lines and FFT line features. The angles in (**a**) are the output angles of the Hough transform, which translate to the angles of the cross-hatching lines since they are perpendicular to the lines in the FFT. The medians shown in the legends in (**a**) and (**b**) where calculated from the angle distribution shown in (**b**), and simply rotated by 90° to match those of the Hough transform output

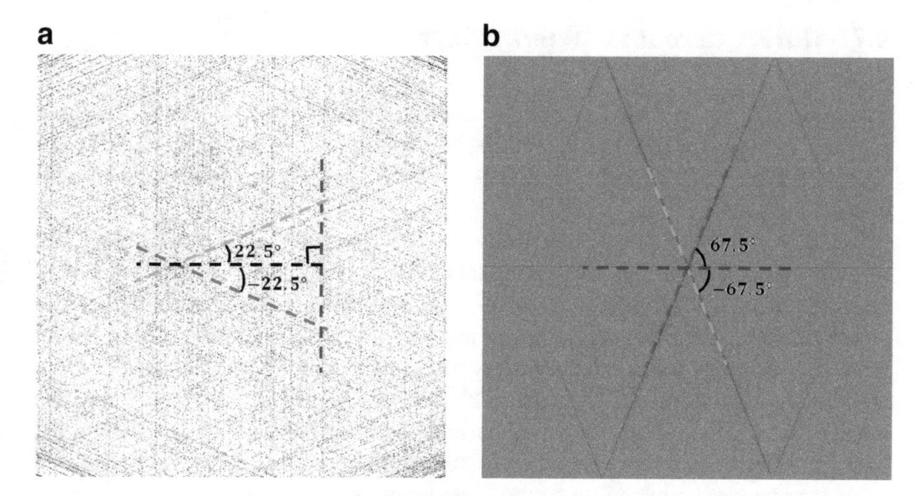

Fig. 4.11 This is the same figure as Fig. 4.7, only with overlaid lines with the angles determined by the Hough transform. The angles of the lines in (**a**) and (**b**) here are those from Fig. 4.10a and b respectively

5 mV, so the initial dark current and actual applied bias (well depth) are used. In those cases, the dark current does not change considerably with bias since they are most likely dominated by diffusion or G-R dark currents.

When the dominating dark current components are tunneling currents, it is important to show how the slope in the dark current SUTR curve changes as the diode debiases. When the curve is highly non-linear, the initial dark current is approximated using only the first two points of the SUTR curve. Following the initial two signal *vs.* time points, three points are used to approximate the instantaneous slope, while the bias corresponds to the mean bias of the points used to approximate the dark current. When a linear behavior in the SUTR curve is reached, a larger number of data points are used to approximate the dark current to reduce the noise since the accumulated dark signal between samples can get close to the read noise level.

A full I-V curve of a given pixel is formed by concatenating the individual I-V curves made from different dark SUTR curves taken with different applied biases, and at the same stable temperature. In cases where the dark signal *vs.* time curve debiases less than 5 mV and is linear throughout the integration from small dark currents due to diffusion and/or G-R, only the initial dark current and well depth are used.

4.7 Dark Current *vs.* Temperature

In addition to the I-V curve, dark current *vs.* temperature (I-T) curves are used together to compare the dark current behavior observed in this array with theory of several dark current mechanisms such as diffusion, G-R, band-to-band and trap-to-band tunneling.

I-T data consist of two different data sets: (1) the initial dark current per pixel which was obtained at the stable temperatures reported in the operability tables in Sects. 5.1.4.1–5.1.4.4, and (2) the warm-up data taken when the liquid helium in our dewar runs out. Four full array frames are read after resetting the array in SUTR mode, immediately followed by an array reset and reading four sub-array (32 rows, all columns) frames in SUTR mode. This data-taking process continued until the temperature reached 77 K. The dark current was then obtained by subtracting the pedestal from the following three frames, taking the average signal from the pedestal subtracted frames and dividing by the average integration time.

The integration time between the full array data frames is 5.8 s, and 0.2 s for the sub-array frames. At lower temperatures, the amount of charge collected by pixels in 0.2 s is on the order of the read-noise ($\sim$23 e^-); we therefore use the 5.8 s integration time data to form the I-T curve at these lower temperatures. As the temperature increases, so does the dark current, saturating the full array frames at a temperature of $\sim$40 K. At a point before the full frames saturate we used the sub-array frames which saturate at a higher temperature ($\sim$50 K) because of the shorter integration time.

Chapter 5
Phase I Results: 13 μm Cutoff Wavelength Devices

The results of the calibration, characterization, and dark current model fit to the LW13 arrays are presented in this chapter. Table 5.1 shows the cutoff wavelength and QE of the four arrays TIS delivered to UR for the first phase of this project. The target cutoff wavelength of these arrays is 13 μm, and the four arrays have three different structure designs. The standard design arrays have the same design as the LW10 detector arrays for the proposed NEOCam mission, only extrapolated to 13 μm. Both H1RG-18369 and H1RG-18509 use TIS proprietary experimental structures designed to reduce tunneling dark currents, designated as design 1 and design 2 (implementing the structure modification ordered by the University of Rochester) respectively in Table 5.1.

TIS provided measurements of the QE without anti-reflection coating and cutoff wavelength from mini-arrays on process evaluation chips (PECs) manufactured with the arrays, at a temperature of 30 K (Table 5.1). The QE as a function of wavelength was measured at UR using a circular variable filter, and we only quote the PEC QE measurements since we found a reasonable agreement with TIS measurements.

5.1 Characterization

To characterize the performance of the LW13 arrays we measured the dark current and well depth per pixel at temperatures ranging from 28 to 36 K with applied reverse bias of 150, 250, and 350 mV, as well as the read-noise per pixel at a temperature of 30 K for two of the arrays. The node capacitance and signal linearity have also been measured at a temperature of 30 K with applied reverse bias of 150, 250, and 350 mV to calibrate the arrays. Most of the characterization results will be covered here by topic, except for the dark current and well depth (operability), which will be discussed separately per array since the results are significantly different.

M. Cabrera, *Development of 15 Micron Cutoff Wavelength HgCdTe Detector Arrays for Astronomy*, Springer Theses, https://doi.org/10.1007/978-3-030-54241-2_5

Table 5.1 Cutoff wavelength and QE for all four LW13 arrays. QE values are expected to increase if arrays had anti-reflection coating

Detector H1RG-	Wafer 2-	Lot-split	Cutoff wavelength (μm)	QE (6–10 μm)
18367	3757	Standard	12.8	74%
18508	3755	Standard	12.7	73%
18369	3763	Design 1	12.4	72%
18509	3759	Design 2	12.6	73%

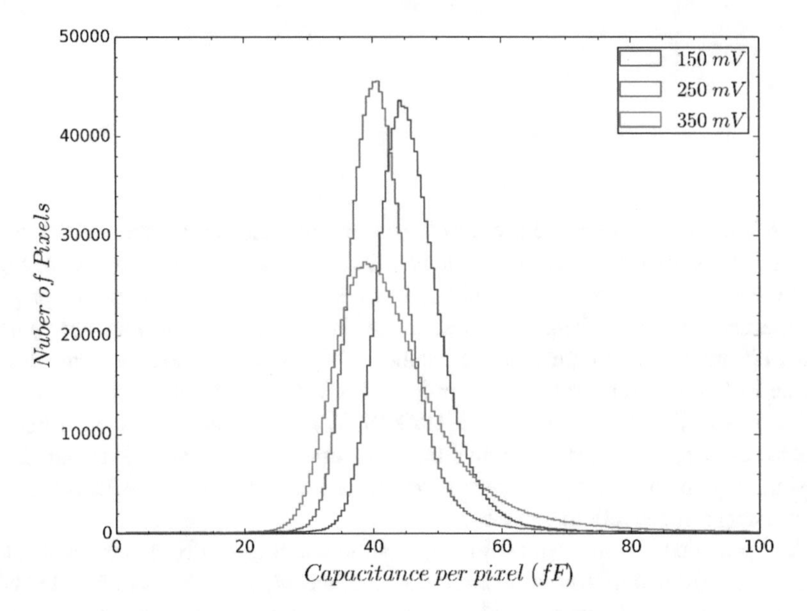

Fig. 5.1 Capacitance per pixel distribution for H1RG-18508 at a temperature of 30 K and at all tested applied biases. These capacitances have not yet been corrected for interpixel capacitance. The spread in the capacitance distribution is primarily due to uncertainty in our probabilistic determination. One standard deviation away from the mean for the 150 and 250 mV distributions is equal to 6.8 and 10.5 fF for the 350 mV distribution

5.1.1 Calibration

The first step in the calibration process is to measure the source-follower FET gain of signal in the multiplexer, and it was measured to be ~0.9 for all four arrays, allowing us to convert between the output-referred and input-referred signal. The data set to measure the source-follower FET gain looks very similar for all four arrays, as that of H1RG-18508 shown in Fig. 4.1. Next, to convert from volts to electrons, the nodal capacitance was measured per pixel at a temperature of 30 K. Figure 5.1 shows the nodal capacitance (not yet corrected for IPC) distribution per pixel for H1RG-18508 for three applied biases, where the median capacitance is then used to convert between the measured ADUs and the signal in electrons for the entire array (see Sect. 4.2.2.2).

Table 5.2 The multiplexer (source-follower FET) gain, IPC coupling parameter α, and the median IPC corrected capacitance in femtofarads for each of the applied biases is given for each of the LW13 arrays

			Median capacitance (fF) Corrected for IPC		
Detector H1RG-	Mux gain	α (%)	150 mV	250 mV	350 mV
18367	0.9	1.04	43	37	35
18508	0.9	1.03	42	38	38
18369	0.9	1.04	38	36	35
18509	0.9	1.12	39	37	34

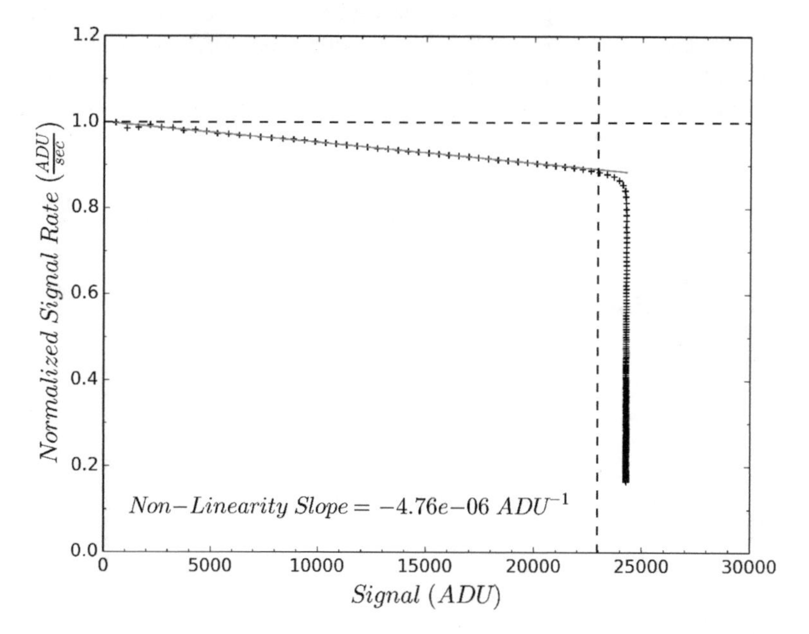

Fig. 5.2 Non-linearity curve obtained from a 50X50 box average of pixels in H1RG-18509 with an applied bias of 150 mV at a temperature of 30 K. The collected signal rate was normalized to that corresponding to the lowest signal. The difference between a value of unity (horizontal dashed line) and the non-linearity curve is the fraction of the signal not observed due to the debiasing of the device. The vertical dashed line corresponds to the mean saturating signal (well depth) for the entire device, which occurs at actual zero bias. The saturation of the non-linearity curve occurs at a higher signal value due to the forward biasing of the detector by the signal flux used to saturate the array

The capacitance then has to be corrected for IPC, determined through the nearest neighbor method. The nodal capacitance is then multiplied by a factor of $1 - 8\alpha$ to correct for IPC (see Sect. 4.2.3). The IPC coupling factor α and corrected capacitance for all four arrays are shown in Table 5.2 for three applied biases.

Once the reset voltage is applied, each pixel debiases with collected charges, resulting in an increasing capacitance from the photo-diode. Figure 5.2 shows the non-linearity curve for H1RG-18509 with an applied bias of 150 mV at 30 K, normalized to the first sampled signal. The normalized signal rate is also referred to as C_0/C, where C_0 is the nodal capacitance at zero collected signal and C is the nodal capacitance. To correct for the varying capacitance, the slope of a fitted line to the non-linearity curve (between 20 and 80% of the saturating signal) is used. Correcting for the variable capacitance can be difficult when dark currents are on the order of the collected signal, which can occur at biases larger than 150 mV. A further step is required at high bias, where the extrapolated line fit is normalized to unity at zero signal.

Figure 5.3 shows the non-linearity curve for H1RG-18369 with an applied bias of 250 mV, where at low signals the rate at which the diode debiases is much faster than is expected from a constant flux illumination. This quick debiasing is due to elevated dark current levels at high bias attributed to quantum mechanical tunneling. The effects of tunneling dark currents which are dominant at larger bias are further discussed in Sect. 5.1.4.

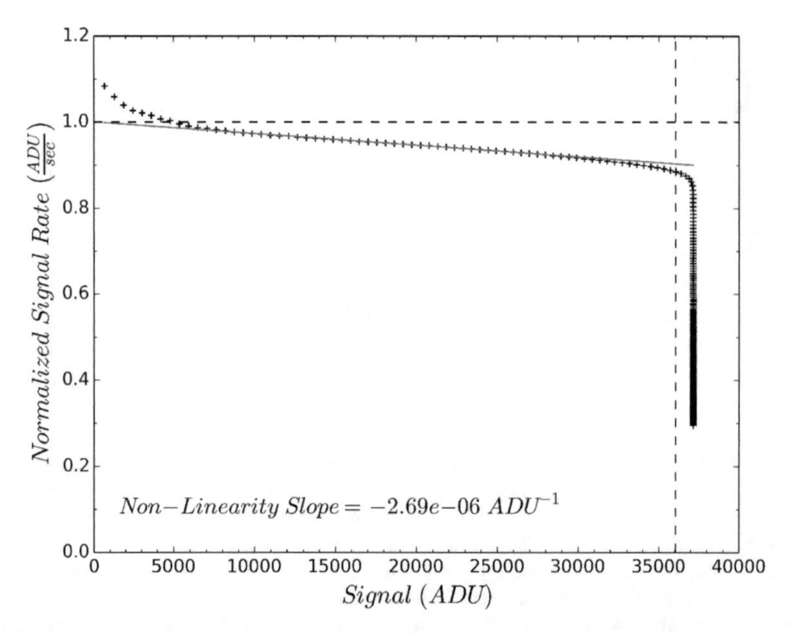

Fig. 5.3 Non-linearity curve obtained from a 50×50 box average of pixels in H1RG-18369 with an applied bias of 250 mV at a temperature of 30 K. The signal rate was normalized and shifted such that the fitted line has a y-intercept of unity. The difference between a value of unity (horizontal dashed line) and the non-linearity curve represents the increase in capacitance, and concomitantly the e^{-}/ADU calibration as the diodes debias. The vertical dashed line corresponds to the signal beyond which the photometric calibration becomes too uncertain for high precision astronomy. The rapid debiasing of the detector at low signals is due to tunneling dark currents

5.1.2 Multiplexer Glow

In multiple occasions during the characterization of H1RG-18367 and H1RG-18369, a uniform elevated current was measured when taking data in the dark. This elevated current affected dark current at 150 mV (34 and 35 K), 250 mV (33–35 K), and read noise at 30 K and 150 mV of applied bias for H1RG-18369. All of the dark current measurements for H1RG-18367 were affected by this elevated current.

A light leak in the test dewar has been ruled out because the anomalous dark current was not present in all data sets obtained under the same conditions; that is to say, there is the possibility that the light leak in the dewar can change if the dewar is opened to change any configurations or the array to be tested, but the dewar remained closed throughout the testing of both arrays when the intermittent elevated current was measured. Figure 5.28 shows an instance where an elevated current affected the dark current and well depth data but was not present in the warm-up data (discussed in Sect. 4.7) at a similar temperature and same applied bias. Dark current has also been ruled out since the dark current mechanisms studied here do not increase by a factor of 10 or more by increasing the temperature by 1 or 2 K for H1RG-18367 and -18369 (see Fig. 1.7 or Tables 5.4 and 5.8).

Therefore, a light source inside of the dewar would have to be responsible for this elevated current. A glow from the two resistors used to control the temperature would be spatially distinguishable, and would not be uniform across the array. The only other source drawing current in the inner sanctum that could cause this glow array-wide is the multiplexer. We believe this current could be a glow from the unit cell FETs in the multiplexer. This intermittent elevated current will be referred to as "mux glow" hereafter when describing the data that were affected. Further work is required to evaluate this anomalous behavior. Significant mux glow from the last row of pixels in the arrays in the four IRAC cameras on the Spitzer Space Telescope were present because the FETs in the last row were purposely left 'on' during long integrations[16].

5.1.3 CDS Read Noise

The total noise per pixel was measured for two arrays, H1RG-18369 and H1RG-18509 at a temperature of 30 K and an applied bias of 150 mV. The read noise was measured in the dark with the cold dark shutter in place. Under these conditions, the CDS total noise is expected to be dominated by the read noise, since the dark current is minimized. If data were obtained under illumination, the shot noise from the photons would have to be subtracted in quadrature from the total noise to obtain the read noise. The read noise data set for H1RG-18369 was affected by the "mux glow" described in the previous section.

Figure 5.4 shows the SUTR set used to measure the dark current and the read noise for the same pixel in H1RG-18369 under the same temperature, bias, and

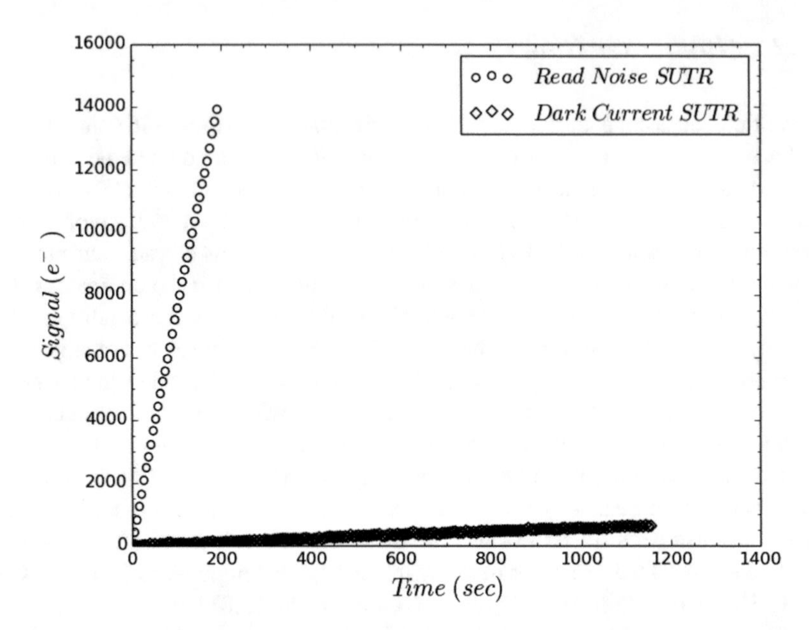

Fig. 5.4 Signal *vs.* time data obtained to measure read noise (only one of the 64 SUTR sets is shown) and dark current for pixel [244,59] in H1RG-18369. Both data sets were obtained under the same conditions: temperature of 30 K, 150 mV of applied reverse bias, and with the cold dark shutter in place to block illumination from the black body source external to the dewar. The only difference between the two sets is a different integration time, 5.5 and 5.8 s for the read noise and dark current data set respectively. The slope in the dark current SUTR data set is 0.5 e^-/s, while the elevated signal collection rate in the read noise SUTR is ~80 e^-/s. This elevated signal is believed to be due to a glow from the multiplexer

with the cold dark shutter in place. The current measured for this pixel in the CDS read noise data set is ~80 e^-/s, whereas the dark current data set yielded a value of 0.5 e^-/s. The elevated current in the data set used to measure the CDS read noise is most likely due to the "mux glow". Figure 5.5 shows the same data sets for a different pixel, which is affected by tunneling dark currents.[1] Initially, the dark current (423 e^-/s) is larger than the flux from the multiplexer glow, and therefore both curves overlap until the dark current decreases below the glow flux.[2] Since the glow is constant, the collected signal rate approaches the same level as that of the pixel shown in Fig. 5.4. H1RG-18509 did not exhibit noticeable "mux glow". Figure 5.6 shows one of the read noise SUTR sets, and the dark current SUTR data set obtained under the same condition for one of the pixels, where both sets are indistinguishable.

[1]Most likely trap-to-band tunneling since band-to-band tunneling does not have this large a contribution at this bias. See Sects. 5.1.4 and 5.2.2 for further discussion on the effects of tunneling dark currents.

[2]For a debiasing diode, tunneling dark currents will decrease over time as charges are collected.

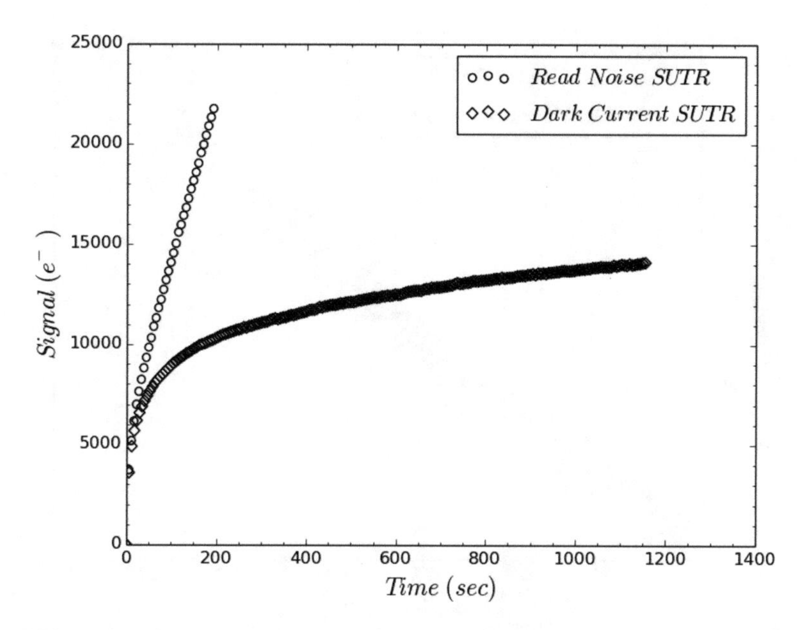

Fig. 5.5 Signal *vs.* time data obtained to measure read noise and dark current for pixel [186,749] in H1RG-18369. The slope in the dark current SUTR data set is 423 e^-/s. Initially, the signal is dominated by tunneling dark currents for both. As the diode debiases, the dark current appears to settle to a linear behavior in the dark current SUTR data set, whereas the read noise SUTR data set settles to the same signal collection rate as the pixel shown in Fig. 5.4 from the mux glow. The dark current behavior will be further discussed in Sect. 5.1.4

Shot noise from the elevated current in H1RG-18369 in this read noise data set is subtracted in quadrature from the total CDS noise to obtain the read noise. The current per pixel is calculated by fitting a line to each of the 64 SUTR curves obtained in the dark to measure the read noise, where the mean of the 64 fitted slopes per pixel is the current used to subtract the shot noise due to the glow from the total noise.

The median elevated current for the entire array H1RG-18369 in the read noise data set was 80 e^-/s, compared to the 0.7 e^-/s measured in the data set obtained to determine the dark current and well depth at the same temperature and bias (discussed in Sect. 5.1.4). The median dark current measured for H1RG-18509 for both data sets are very similar, 0.8 e^-/s in the read noise data set, and 0.7 e^-/s in the dark current and well depth data set.

Figure 5.7 shows the distribution of rms read noise per pixel for both arrays, where the median noise is 22.9 e^- and 23.2 e^- for H1RG-18369 and H1RG-18509 respectively. The majority of pixels with low noise approximately between 10 and 15 e^- in the read noise distribution for H1RG-18509 and in the total noise distribution for H1RG-18369 are located in the "picture frame" region[47] of the arrays. Figure 5.8 shows a binarized map of the rms read noise, where pixels with read noise values between 10 and 15 e^- are shown in black. This map shows the

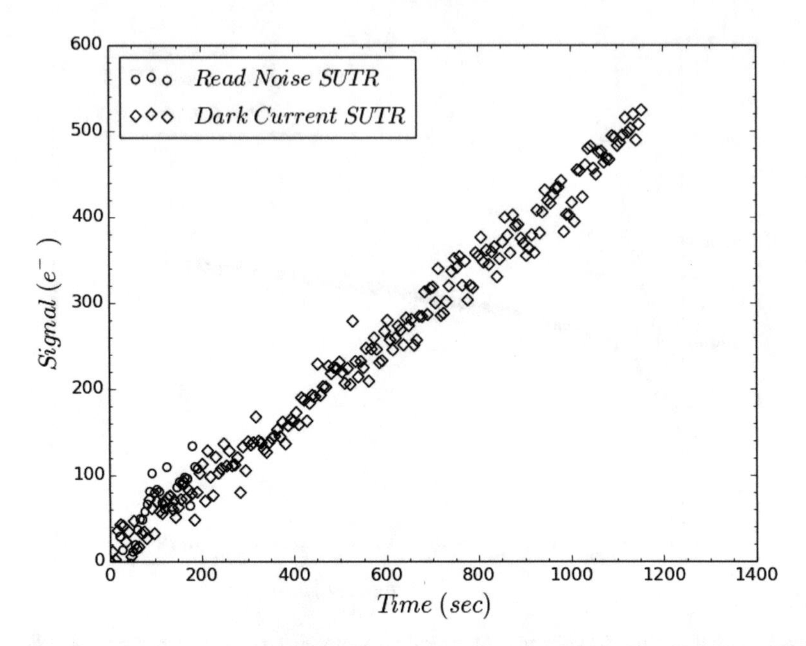

Fig. 5.6 Signal *vs.* time data obtained to measure read noise and dark current for pixel [836,453] in H1RG-18509. Both data sets were obtained under the same conditions: temperature of 30 K, 150 mV of applied reverse bias, and with the cold dark shutter in place. The slope in the dark current SUTR data set is 0.4 e^-/s. Unlike the read noise data set for H1RG-18369, mux glow was not present in the H1RG-18509 data; hence, both curves overlap since the signal collection is dominated by dark current in both instances

spatial distribution of the small peak in the rms noise distribution from Fig. 5.7 in the same range. The vast majority of these pixels are concentrated in the picture frame of the arrays. The picture frame (a shift in voltage offset) can be observed in all individual frames obtained in SUTR mode, and it subtracts exactly when the pedestal is subtracted from the rest of the images.

The median read noise measured for the LW10 arrays has recently been shown to be 19 e^-[24], very close to that calculated for these two LW13 arrays. Since read noise is not an issue with these arrays, and the focus of this project is to evaluate the performance of the detector material (read noise determined by mux), the operability is determined by the dark current and well depth.

5.1.4 Dark Current and Well Depth (Operability)

As was mentioned in Chap. 1, the pixel operability requirements for the proposed NEOCam mission LW10 devices are QE (>55%), read noise <36 e^-, dark currents <200 e^-/s, and well depths >46 ke^-. In addition, the NEOCam mission requires that each array should have >90% of operable pixels at a temperature of 40 K. Since

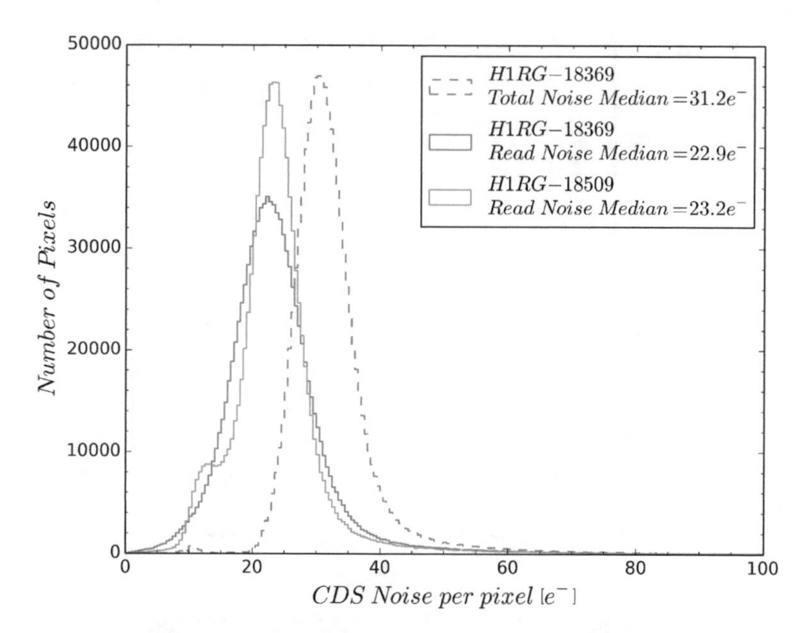

Fig. 5.7 Histogram of CDS read noise per pixel for H1RG-18369 and H1RG-18509 at a temperature of 30 K and applied bias of 150 mV. The CDS read noise was obtained by subtracting the contribution from dark current or a signal flux from the total noise in a CDS image. The total noise is plotted for H1RG-18369 to show the contribution from the "mux glow" to the noise of the detector. The dark current contribution from H1RG-18509 to the total noise was minimal, and that array did not exhibit a "mux glow"; therefore, we only present the read noise distribution for this array

this is a developmental project to determine the prospect of using HgCdTe detector arrays at these extended cutoff wavelengths instead of Si:As arrays, there are no set operability requirements.

The focus of the first phase of this project (LW13 arrays) was to determine the best array design to move forward with final goal of reaching the 15 μm cutoff wavelength. Therefore, we imposed operability requirements similar to those of the NEOCam mission to compare the performance among the different array designs. The QE measurements from TIS were well above the NEOCam requirement, and they are expected to increase to above 90% if the arrays are AR-coated. The median read noise results for two of the arrays, which had two of the three array designs, had median values of noise below the LW10 NEOCam array requirement. Since we're primarily interested in the infrared detectors themselves, the operability here is focused on the dark current and well depth.

Dark current and well depth for the LW13 arrays are measured at temperatures ranging from 28 to 36 K, and applied reverse bias of 150, 250, and 350 mV. At all temperatures, a 200 e^{-}/s operability dark current requirement is imposed on the pixels, while the well depth requirement increases with increasing applied bias. In this section, the effects of dark current on well depth and operability are discussed.

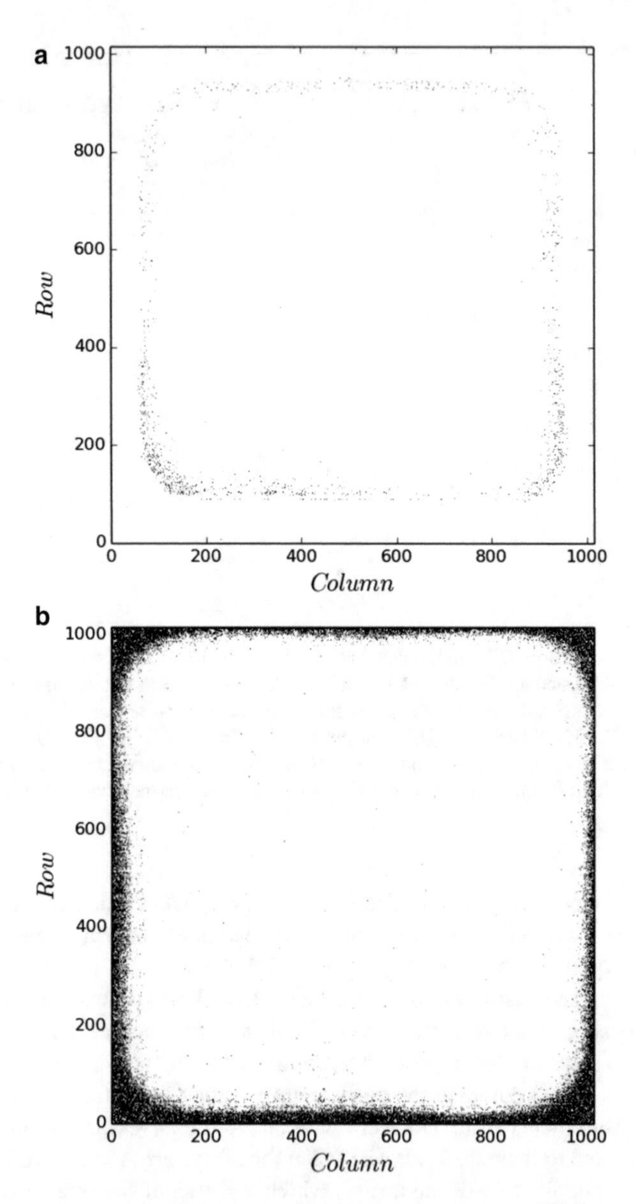

Fig. 5.8 Binarized map of the read noise for both H1RG-18369 (**a**) and H1RG-18509 (**b**). Pixels with read noise between 10 and 15 e^{-} at a temperature of 30 K are shown in black. This shows the spatial distribution of the small peak in that range of the rms read noise distribution of the arrays. These pixels are located on the picture frame of their respective arrays

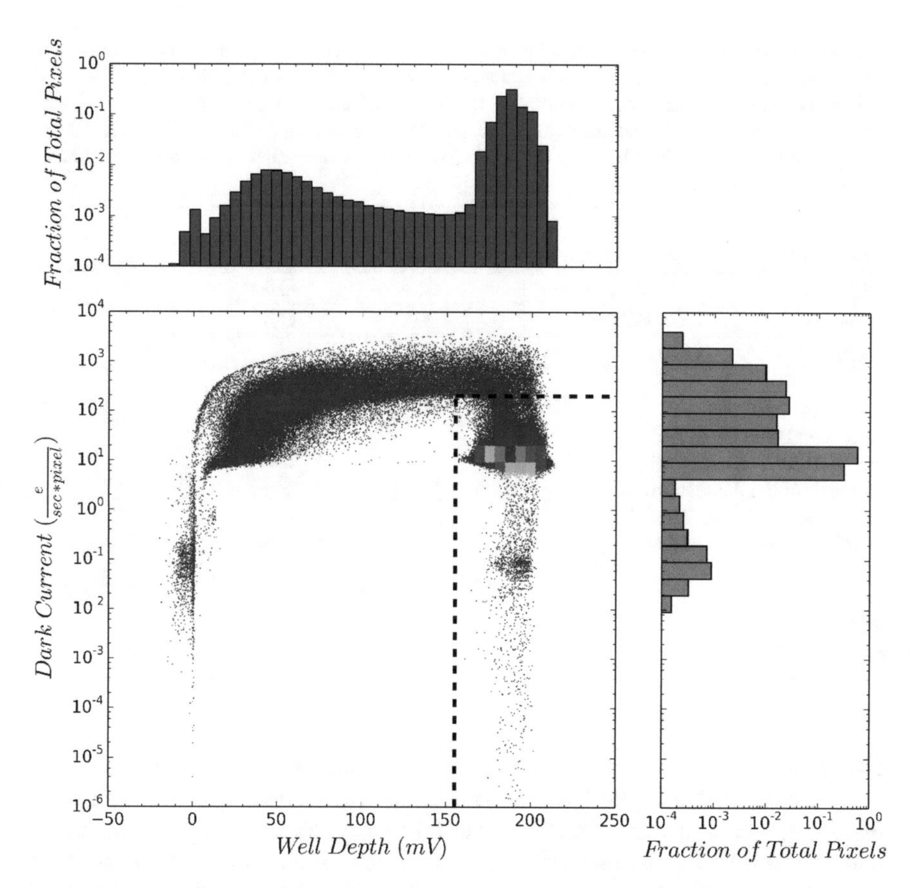

Fig. 5.9 Current in the dark *vs.* well depth distribution per pixel for H1RG-18367 at a temperature of 28 K and an applied bias of 150 mV, showing operable pixels below and to the right of the dashed line. The well depth requirement of 155 mV ($\sim$41 ke^-) was chosen to include the majority of the good pixels, and we kept this requirement for higher temperatures with the same applied bias. Note that the well depth corresponds to the actual initial detector bias, ranging from 155 to 215 mV at the beginning of the integration among the operable pixels for this data set

In addition to showing the operability at different temperatures and biases, tables with the median dark current and well depth are also presented.

5.1.4.1 H1RG-18367

Figure 5.9 shows the distribution of initial dark current *vs.* well depth per pixel for H1RG-18367 at a temperature of 28 K and applied bias of 150 mV, where the operable pixels are below and to the right of the dashed lines. The dark current distribution per pixel for this array peaks around 10 e^-/s for an applied bias of 150 mV and a temperature of 28 K: we attribute this dark current to the "mux glow"

Table 5.3 Operability pixel percentage for H1RG-18367 with different applied biases and temperatures. Operability requirements include currents in the dark (A glow from the mux affected all of the dark current measurements for this array. This "mux glow" was highly variable, with median current of approximately 10, 130, and 156 e^{-}/s at temperatures of 28, 30, and 32 K respectively with an applied bias of 150 mV.) below 200 e^{-}/s and well depths greater than those indicated in the table

		Operability (%)		
Applied bias	Well depth minimum	T=28 K	T=30 K	T=32 K
150 mV	155 mV (∼41 ke^{-})	91.4	90.9	88.9
250 mV	255 mV (∼59 ke^{-})	90.1	89.5	88.2
350 mV	310 mV (∼67 ke^{-})	1.4	0.5	3.9

Table 5.4 Median dark current and well depth for H1RG-18367. "Mux glow" was present in all dark current measurements. With 350 mV of bias, the decrease of dark current with increasing temperature (due to increasing the band gap energy) is indicative of tunneling dark currents

	Median dark current (e^{-}/s) Median well depth (ke^{-}, mV)		
Applied bias	T=28 K	T=30 K	T=32 K
150 mV	10	130	156
	50, 186	50, 185	48, 180
250 mV	12	116	139
	67, 287	67, 286	66, 281
350 mV	379	359	321
	82, 379	82, 379	82, 376

discussed in Sect. 5.1.2 given its uniformity across the entire array, and lack of bias dependence. The "mux glow" was also observed at temperatures of 30 and 32 K for this array, with a median "dark" current of 130 and 156 e^{-}/s respectively with a bias of 150 mV. This accounts for some of the loss of operability at 30 and 32 K *c.f.* 28 K. The percentage of operable pixels for different temperatures and applied biases are reported in Table 5.3, along with the well depth requirements. Median dark current and well depth measured for this array are shown in Table 5.4.

Figure 5.10 shows the discharge history of a pixel in the dark at three different applied biases for this device. The initial larger curvature observed in the signal with an initial applied bias of 350 mV is a consequence of quantum tunneling dark currents which are exponentially dependent on bias. As the pixel debiases, tunneling dark currents decrease and the dark current approaches the constant behavior with bias expected from steady "mux glow". The 250 mV curve shows a smaller curvature at the beginning of the ramp, indicating that there is a small contribution from the tunneling dark currents, but the curve appears to be mostly dominated by the glow from the multiplexer.

The major limiting factor in the operability of these arrays at larger bias and low temperatures has been tunneling dark currents, and it is much more evident at biases of 350 mV since trap-to-band and especially band-to-band tunneling are strongly dependent on the detector back bias [17, 18, 21]. At higher bias, most of the pixels have dark currents exceeding 200 e^{-}/s, giving very low operabilities. The

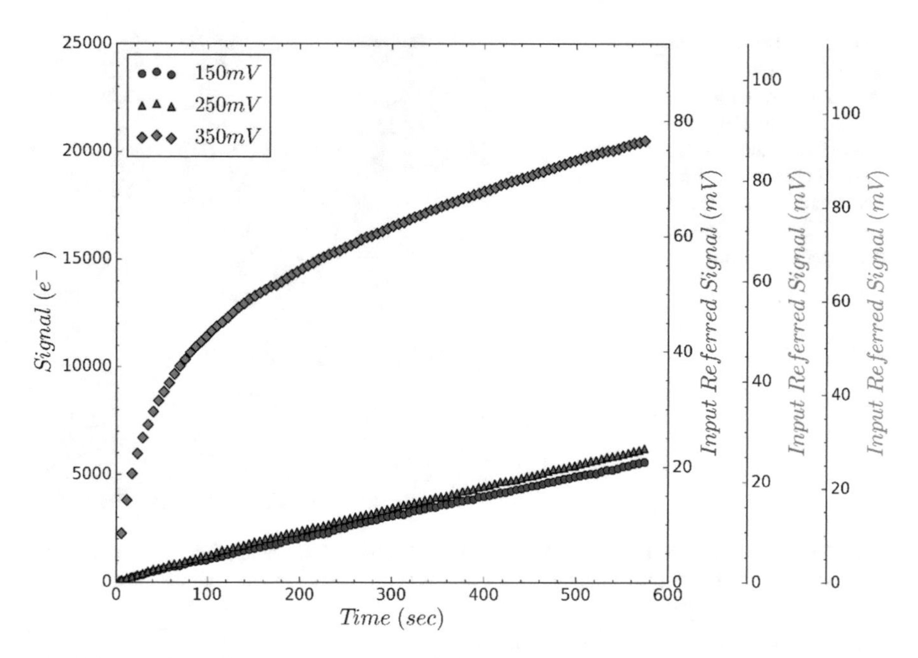

Fig. 5.10 Discharge history for pixel [532, 468] from H1RG-18367 at a temperature of 28 K and at applied reverse bias of 150, 250, and 350 mV in the dark. On the left y-axis scale we show the signal in electrons, while the right y-scale axis shows the input referred signal in mV. The three different scales on the right correspond to the different diode capacitance measured (from left to right) at 150, 250, and 350 mV of applied biases. This pixel is inoperable at 350 mV of applied bias, since the initial dark current exceeds the 200 e^-/s cutoff

trap-to-band dark current is highly variable from pixel to pixel, but the band-to-band tunneling affects all of the pixels in the array equally.

The inoperability at low temperatures and low detector bias (~28 K and 150 mV respectively) appears to be dominated by trap-to-band tunneling currents, and not by band-to-band tunneling. If band-to-band tunneling was responsible for the inoperability at low biases, all pixels would be inoperable since it affects all pixels uniformly. Rather, the effect of trap-to-band tunneling can be seen in the operability map for all four arrays (Fig. 5.11 for H1RG-18367) where a set of high dark current and/or low well depth, hence inoperable pixels form the cross-hatching pattern discussed in Sect. 4.5. The weakness of the central region on the FFT indicates little array-wide excess dark current, as would result from band-to-band tunneling, G-R, and diffusion currents. The pixels that lie along the cross-hatching pattern will suffer from trap-assisted dark currents. The vast majority of inoperable pixels show a large curvature at the beginning of the dark SUTR curve (see Sect. 5.1.5). The curvature in the signal collection at this low bias can only be explained by trap-to-band tunneling, and not G-R. Dark SUTR examples for pixels dominated by trap-to-band and band-to-band are shown for array H1RG-18508 in the following

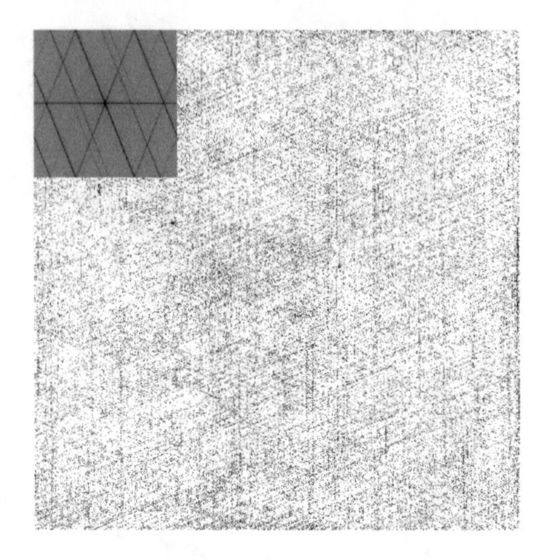

Fig. 5.11 Operability map for H1RG-18367 at a temperature of 28 K and applied bias of 150 mV, where inoperable pixels are shown in black. The (log magnitude) FFT of the operability map is shown in the upper-left corner

Table 5.5 Operability pixel percentage for H1RG-18508 with different applied biases and temperatures. Operability requirements include dark currents below 200 e^-/s and well depths greater than those indicated in the table

Applied bias	Well depth minimum	Operability (%)						
		T=28 K	T=30 K	T=32 K	T=33 K	T=34 K	T=35 K	T=36 K
150 mV	155 mV (∼41 ke^-)	93.7	93.5	93.2	92.9	92.0	61.5	63.7
250 mV	245 mV (∼58 ke^-)	89.4	91.5	91.4	91.0	89.9	81.1	67.2

section, where similar trends are true for all four arrays. The results of the direction of the cross-hatching pattern in all four LW13 arrays are discussed in Sect. 5.1.5.

5.1.4.2 H1RG-18508

H1RG-18508 has a higher operability (Table 5.5) at higher temperatures than its counterpart H1RG-18367 (both were grown and processed in the same standard way, and both have similar cutoff wavelengths), and "mux glow" was not detected for this array. The dark current *vs.* well depth distribution for this array at 28 K and an applied bias of 150 mV is shown in Fig. 5.12, where the majority of pixels had dark currents below 1 e^-/s. This is the typical distribution that is expected from arrays without a "mux glow", where the dark current has a broader distribution. All four LW13 arrays had the lowest dark current at a temperature of 28 K and 150 mV of applied bias. Median dark current and well depths for H1RG-18508 are presented in Table 5.6.

The operabilities at 350 mV of applied bias are below 1% at all temperatures, and have therefore been omitted from the operability table. The inoperability at larger applied bias is evidence of tunneling dark currents. The large curvature in the dark signal *vs.* time curve is observed with increasing bias in Fig. 5.13, which is a

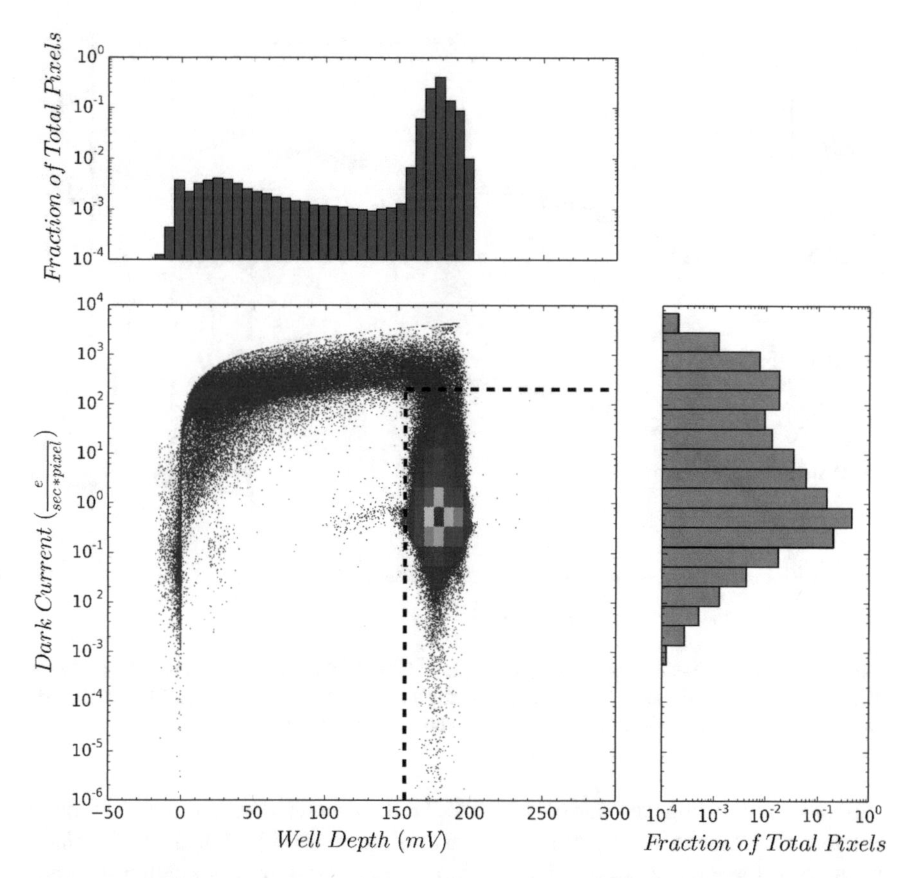

Fig. 5.12 Dark current *vs.* well depth distribution per pixel for H1RG-18508 with an applied bias of 150 mV at a temperature of 28 K. Operable pixels lie below and to the right of the dashed line. The detector bias at the beginning of integration is similar to that of array H1RG-18367 (*c.f.* Fig. 5.9)

Table 5.6 Median dark current and well depth for H1RG-18508

	Median dark current (e^-/s)						
	Median well depth (ke^-, mV)						
Applied bias	T=28 K	T=30 K	T=32 K	T=33 K	T=34 K	T=35 K	T=36 K
150 mV	0.5	1.4	6	14	32	65	143
	46, 177	45, 173	45, 171	44, 170	44, 169	41, 157	42, 160
250 mV	57	39	34	39	51	71	144
	65, 275	64, 271	63, 269	63, 268	63, 267	60, 254	60, 253
350 mV	780	764	730	713	693	510	279
	73, 328	74, 331	76, 337	76, 341	77, 344	76, 341	74, 331

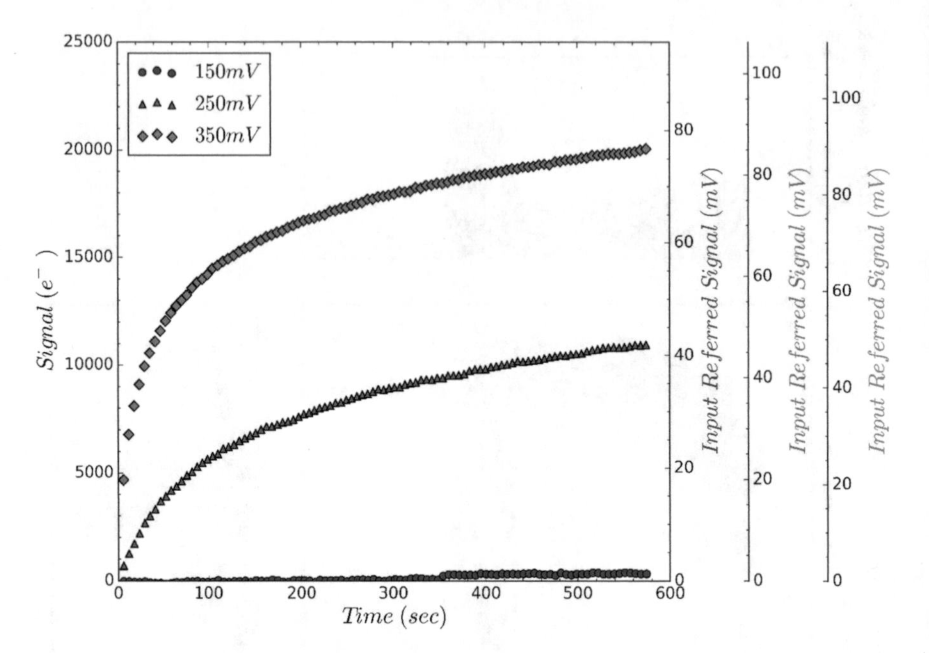

Fig. 5.13 Discharge history for pixel [543, 424] from H1RG-18508 at a temperature of 28 K in the dark. The initial dark currents (and well depths) for three SUTR curves at 150, 250, and 350 mV are: 0.4 e^-/s (181 mV), 120 e^-/s (281 mV), and 807 e^-/s (326 mV) respectively

hallmark of tunneling dark currents. The SUTR data obtained with an applied bias of 150 mV are linear as is expected if thermal dark currents (diffusion and G-R, which are a weak function of bias), a light leak, or a mux glow are dominating the dark current as the pixel debiases. Another indication of a strong presence of tunneling dark currents is the decreasing initial dark current with increasing temperature. With increasing temperature, the band gap energy increases, lowering the tunneling probability. For this array, tunneling dark currents are the dominant component, initially after reset, starting with an applied reverse bias of 250 mV. This can be seen in Table 5.6 where the median initial dark current decreases as the temperature increases from 28 to 32 K. Once the dark current begins to increase with increasing temperature, thermal dark currents take over as the principal dark current component or are comparable with the tunneling component. With an applied reverse bias of 350 mV, tunneling currents appear to be the main source of dark current immediately following the reset at all tested temperatures.

5.1.4.3 H1RG-18369

This array has a similar dark current and well depth distribution as H1RG-18508, where the majority of the pixels achieved dark current levels less than 1 e^-/s at a temperature of 28 K and an applied bias of 150 mV. Table 5.7 shows the

Table 5.7 Operability pixel percentage for H1RG-18369 with different applied biases and temperatures. Operability requirements include dark currents below 200 e^{-}/s and well depths greater than those indicated in the table

		Operability (%)					
Applied bias	Well depth minimum	T=28 K	T=30 K	T=32 K	T=33K	T=34 K	T=35 K
150 mV	155 mV ($\sim$37 ke^{-})	93.0	92.8	92.7	92.6	0.0*	0.0*
250 mV	255 mV ($\sim$57 ke^{-})	91.5	91.3	91.1	0.0*	0.0*	0.0*

* The "mux glow" affected the noted operabilities for this array, increasing the median dark current by a factor of $\sim$100 when the temperature was increased from 33 to 34 K with an applied bias of 150 mV, and by a factor of $\sim$43 when increasing the temperature from 32 to 33 K with an applied bias of 250 mV, effectively making the entire array inoperable.

Table 5.8 Median dark current and well depth for H1RG-18369. Dark currents marked with an asterisk were affected by a "mux glow"

	Median dark current (e^{-}/s) Median well depth (ke^{-}, mV)					
Applied bias	T=28 K	T=30 K	T=32 K	T=33 K	T=34 K	T=35 K
150 mV	0.5	0.7	2.4	5	481*	698*
	43, 182	42, 177	41, 174	43, 181	41, 173	40, 169
250 mV	24	17	14	604*	600*	791*
	64, 283	63, 278	62, 275	62, 275	62, 273	61, 269
350 mV	730	682	616	989*	930*	957*
	78, 355	78, 357	79, 359	80, 364	80, 364	79, 361

operability for this array, where the unusual drop in operability when increasing the temperature by 1° with an applied bias of 150 and 250 mV is most likely due to "mux glow". Table 5.8 shows the median dark current and well depth for this array, where the increase in median dark current by a factor of $\sim$100 at a temperature of 34 K and 150 mV of bias cannot be explained by any of the known dark current mechanisms. Additionally, the decrease in dark current with increasing temperature, for an applied bias of 350 mV from 28–32 K, is consistent with tunneling dark currents. If the increase in dark current at 33 K was due to thermal currents overtaking tunneling currents, we would expect the dark current to continue to rise with increasing temperature, and this is not the case. Figure 5.14 shows the dark current distribution for this array at 33 and 34 K with an applied bias of 150 mV. The histogram shows a steep cutoff in dark current around $\sim$350 e^{-}/s, which would suggest either a very uniform dark current source, or as has been proposed, a glow from the multiplexer. It will be shown in Sect. 5.2.2 that at 34 K, G-R should be the dominant source of dark current in this array. If G-R currents were responsible for the distribution in Fig. 5.14, a broader distribution in the dark current histogram, extending towards lower values, is expected since this source of dark current is not uniform across the whole array due to its dependence on pixel trap characteristics.

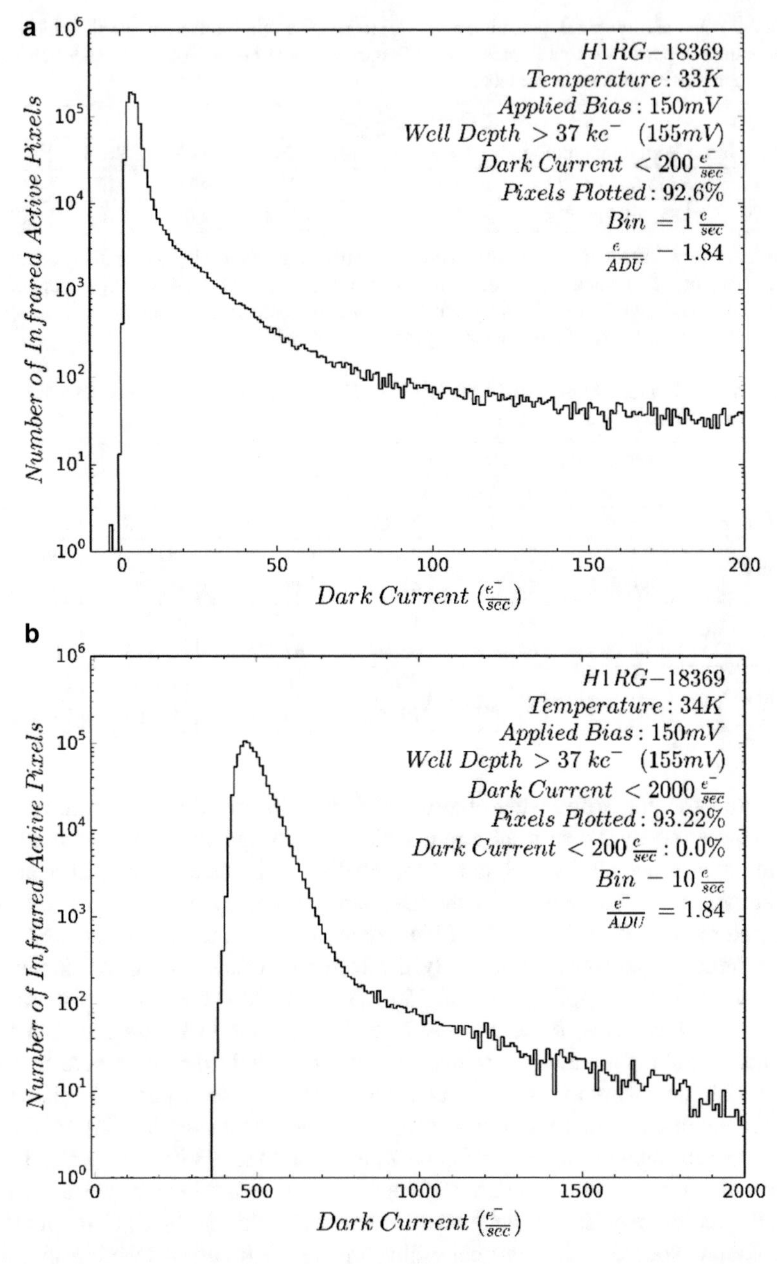

Fig. 5.14 Dark current distribution for pixels in H1RG-18369 with well depths greater than 37 ke^- at temperatures of 33 K (**a**) and 34 K (**b**), both with an applied bias of 150 mV. The increase in current by a factor of ∼100 by increasing the temperature by 1 K is most probably due to variable "mux glow"

Furthermore, despite the experimental design used to reduce tunneling dark currents, this array also exhibited low operabilities at 350 mV (<1%) as did the standard growth arrays (H1RG-18367 and 18508).

5.1.4.4 H1RG-18509

To mitigate the effects of the tunneling dark currents, TIS and UR developed several modifications to its baseline NEOCam design, designated as designs 1 and 2, shown in Table 5.1. H1RG-18509 was from the UR Design-2 lot split and outperformed the other LW13 arrays in terms of operability at all applied biases, but especially at 350 mV, and at higher temperatures (Table 5.9). Table 5.10 shows the median dark current and well depths measured for this array at different temperatures and applied bias. This array has a slightly shorter wavelength cutoff compared with the standard growth arrays, but its superior performance is attributed to the pixel design.

Design 2 had a positive effect on reducing the quantum tunneling currents as is readily apparent in curves of the time dependence of charge collected in the dark, where most of the individual SUTR curves for H1RG-18509 did not exhibit highly curved discharge behavior such as observed for the other arrays. Figure 5.15 shows the histogram of curvature values (initial second derivative of the SUTR curves) per pixel, for all LW13 arrays with an applied bias of 350 mV. The peak of the histogram near zero for H1RG-18509 demonstrates a nearly linear behavior in the charge collected over time for the majority of pixels, in contrast to the behavior

Table 5.9 Operable pixel percentage for H1RG-18509 with different applied biases and temperatures. Operability requirements include dark currents below 200 e^-/s and well depths greater than those indicated in the table

		Operability (%)					
Applied bias	Well depth minimum	T=28 K	T=30 K	T=32 K	T=34 K	T=35 K	T=36 K
150 mV	155 mV (~38 ke^-)	95.9	95.9	95.8	95.7	94.7	85.5
250 mV	255 mV (~59 ke^-)	94.7	94.6	94.5	94.0	92.7	44.7
350 mV	355 mV (~75 ke^-)	90.3	92.2	93.1	92.3	90.8	19.4

Table 5.10 Median dark current and well depth for H1RG-18509

	Median dark current (e^-/s) Median well depth (ke^-, mV)					
Applied bias	T=28 K	T=30 K	T=32 K	T=34 K	T=35 K	T=36 K
150 mV	0.2	0.7	3.5	19	41	82
	44, 182	43, 177	43, 174	42, 174	42, 173	40, 163
250 mV	0.3	0.8	3.8	20	43	76
	65, 283	64, 278	63, 276	63, 275	63, 274	59, 255
350 mV	1.8	1.8	4	19	42	67
	81, 385	80, 380	80, 378	80, 377	80, 376	74, 348

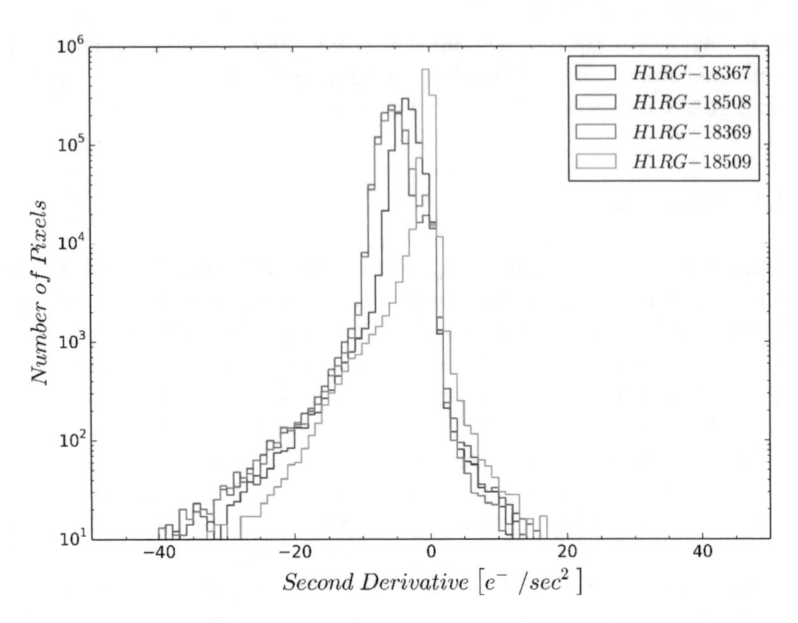

Fig. 5.15 Histogram of the initial curvature from SUTR curves for all LW13 arrays at 28 K and an applied bias of 350 mV

of the other three arrays shown by the peaks of their respective histograms shifted towards larger negative values.

5.1.5 Cross-Hatching (Inoperable Pixels)

A common occurrence among inoperable pixels at low temperatures and low bias is that the majority of those pixels have a curvature in the dark signal *vs.* time curves used to determine the initial dark current, suggesting the presence of trap-to-band tunneling. The dark SUTR curves for five inoperable pixels are plotted in Fig. 5.16. The varying initial discharge behavior of all five pixels is expected from trap-to-band tunneling currents from varying trap characteristics for each of the pixels. As it has already been mentioned multiple times, if band-to-band tunneling was responsible for this behavior, the discharge history of the pixels would be uniform. To show this explicitly without the need to fit any dark current models, Fig. 5.17 shows the dark SUTR curves for five operable pixels. The uniform initial discharge of the five pixels implicitly shows that band-to-band tunneling is the dominant component of dark current.

In the same way that the initial second derivative of the dark SUTR curves was used to demonstrate the reduced tunneling currents in H1RG-18509 (see Sect. 5.1.4.4), Fig. 5.18 shows the initial curvature of operable and inoperable pixels at a temperature of 28 K and applied reverse bias of 150 mV to show that

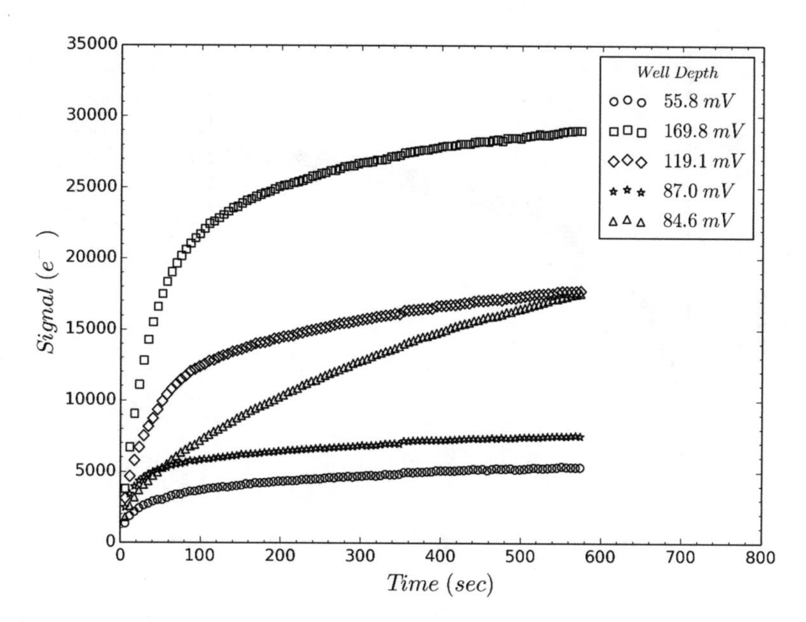

Fig. 5.16 Discharge history for five inoperable pixels (dark current >200 e^-/s and/or well depth <41 ke^-) from H1RG-18508 at a temperature of 28 K and 150 mV of applied reverse bias in the dark. The well depth for each of the pixels is presented in the legend to provide context when interpreting the behavior of the curves. The curve with open circle markers may appear to have the smallest initial dark current from the five pixels shown, but it also has the smallest well depth. Most likely, this pixel actually had the largest dark current immediately following the reset, and debiased considerably before the pedestal frame, appearing to have lower dark current. The large curvature, and the non-uniformity among the behavior of the different pixels is indicative of trap-to-band tunneling current

the majority of inoperable pixels exhibit trap-to-band tunneling. The majority of operable pixels have a linear dark SUTR curve, demonstrated by the peak near zero in the curvature histogram, whereas the inoperable pixels show an initial curvature in the discharge curve denoted by the peak of the histogram towards negative values. Only inoperable pixels with a well depth greater than 50 mV were used to construct Fig. 5.18b in order to show pixels that did not appreciably debias before the pedestal frame.

Impurities or dislocations in the HgCdTe bulk material can be the sources of the energy traps within the band gap of the semiconductor that contribute to trap-assisted dark currents. Though we cannot distinguish between the two cases when fitting the trap-to-band tunneling model to data from these devices, if the cross-hatching pattern seen in these arrays is matched with that expected from HgCdTe grown on CdZnTe substrates, then the argument can be made that the subset of pixels in the cross-hatching pattern most certainly have dislocations (see Sect. 4.5) that give rise to trap-assisted tunneling. Impurities would be randomly distributed across the array.

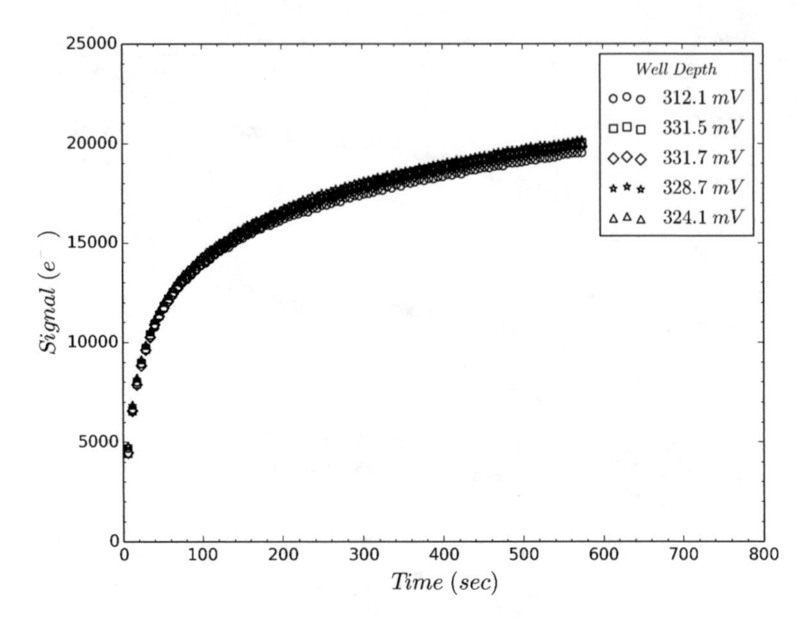

Fig. 5.17 Discharge history for five operable pixels from H1RG-18508 at a temperature of 28 K and 350 mV of applied reverse bias in the dark. This bias was chosen to show the typical uniform behavior of pixels with band-to-band tunneling as the dominant component of dark current initially following the reset

The fast Fourier transform (FFT) of the operability map on the upper left corner of Fig. 5.11 shows the cross-hatching lines distinctly, but rotated by 90°. The angles between the three cross-hatching lines (shown in Fig. 4.4) were calculated using the method described in Sect. 4.5.1 by applying the Hough transform to the FFT of the operability maps for all four LW13 arrays.

Figure 5.19 shows the Hough transform and its input image used to find the angles of the cross-hatching pattern in H1RG-18367. The angles of the Hough transform correspond to the angle of a line perpendicular to the line features in the FFT image. Since the FFT line features are rotated by 90° from the cross-hatching pattern in the operability map, the Hough transform angles match those of the cross-hatching w.r.t. the horizontal (Fig. 5.20).

The cross-hatching pattern angles found in this work, Table 5.11, are in close agreement with the angles found in Martinka et al.[29] and Chang et al.[30] made by the cross-hatching lines parallel to the $[\bar{2}31]$, $[\bar{2}13]$, and $[01\bar{1}]$ directions. The cross-hatching pattern in the operability map of H1RG-18509 is rotated by 90° from that seen in the other three LW13 arrays, but the relative angles between the cross-hatching directions (θ_α and θ_β) are consistent with those found for the rest of the arrays. The wafer used to make H1RG-18509 was most likely simply rotated by 90° when hybridized to the multiplexer. Figures 5.21, 5.22, and 5.23 show the operability map for H1RG-18509 at a temperature of 28 K and applied bias of 150 mV, and the angles found for the cross-hatching pattern in this array.

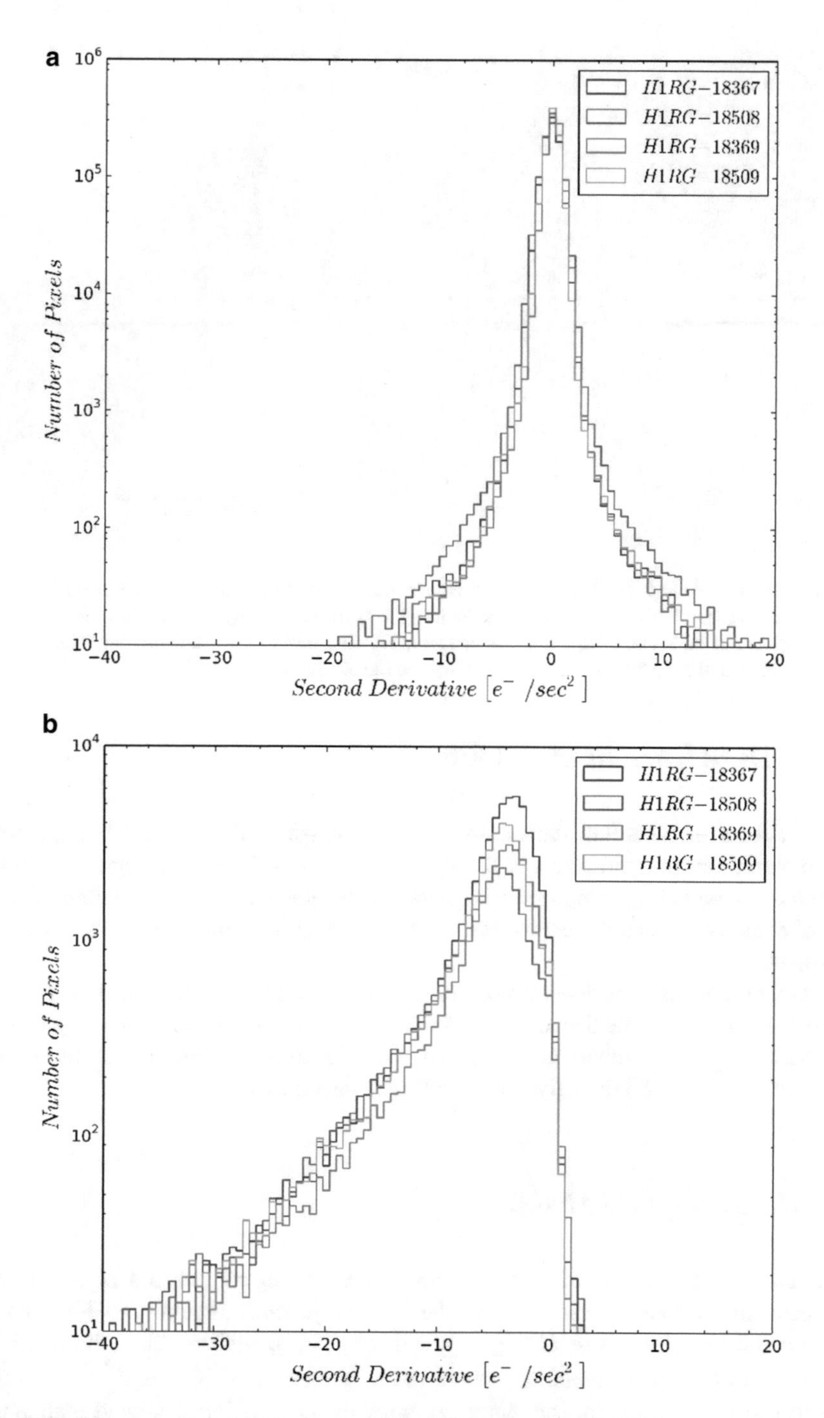

Fig. 5.18 Histogram of the initial curvature from SUTR curves for all LW13 arrays at 28 K and an applied bias of 150 mV for (**a**) operable and (**b**) inoperable pixels

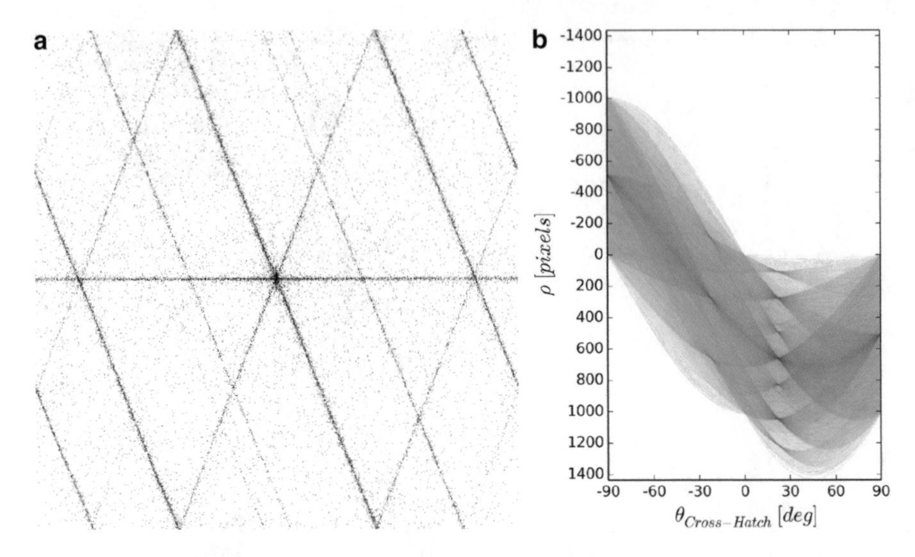

Fig. 5.19 Binarized input image (**a**) and Hough transform (**b**) to determine the cross-hatching angles in H1RG-18367. The input image is the FFT of the operability map for H1RG-18367 at a temperature of 28 K and an applied bias of 150 mV. A threshold of 1.5 standard deviations above the mean was used to binarize the image shown on the top-left corner of Fig. 5.11

5.2 Dark Current Model Fits

The effects of tunneling dark currents were shown in Sects. 5.1.4.1–5.1.5, and comparing our measurements to theory will allow us to assess the degree to which each of the tunneling components is dominant and how we can mitigate these effects further as we continue to extend the cutoff wavelength for the second phase of this project.

Given the different dependences on temperature and bias for the dark current mechanisms, we fit the thermal currents, which are strongly temperature dependent, to dark current *vs.* temperature (I-T) data, while the tunneling currents are fit to dark current *vs.* bias (I-V) data given their strong dependence on bias.

5.2.1 *Theoretical Models*

Equations 3.23 and 3.25 in Sect. 3.3 show that tunneling currents are exponentially dependent on band gap and electric field in the junction region. The I-V curves of several operable pixels at large applied bias (>200 mV) matched the trend of a fit to band-to-band tunneling, where the only parameter to fit is $E_g^{3/2}/E \equiv \beta$. If indeed these pixels (or the majority) are dominated by band-to-band tunneling, this would allow us to get the most accurate estimate of the parameter β since

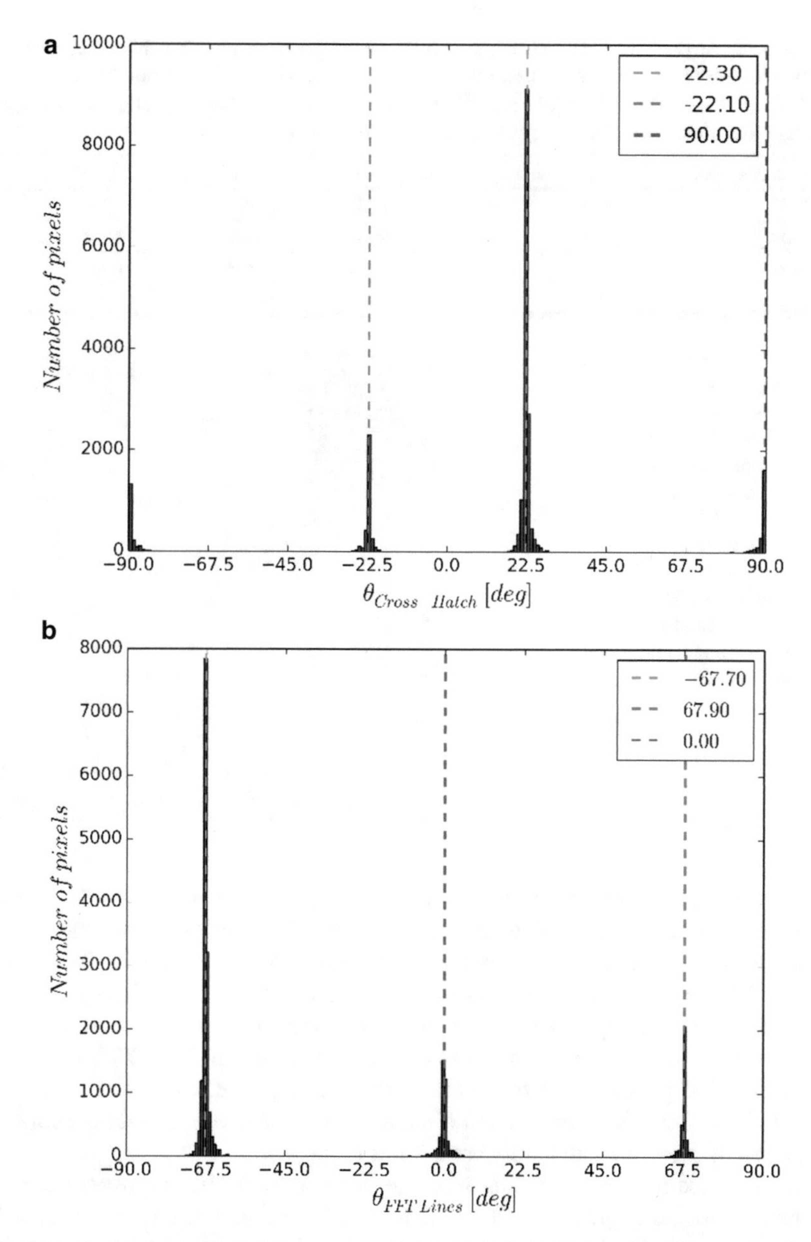

Fig. 5.20 Angle distribution with respect to the horizontal of the cross-hatching pattern in the operability map of H1RG-18367 (**a**) and that corresponding in Fourier space (**b**). The cross-hatching angle (**a**) distribution is obtained from the output of the Hough transform, where this same distribution is simply rotated by 90° to obtain that of the line features in the FFT image of the operability map (used as the input image to the Hough transform)

Table 5.11 Angles (with respect to the horizontal) of the three cross-hatching directions in operability maps. θ_α is the angle between the $[\bar{2}31]$ and $[\bar{2}13]$ directions, while θ_β is the angle made by the intersection of the $[\bar{2}31]$ and $[01\bar{1}]$ directions (shown in Fig. 4.4). The cross-hatching angles reported in Martinka et al. [29] and Chang et al. [30] are θ_α and θ_β

Detector H1RG-	Cross-hatching line angles (Deg.)			θ_α (Deg.)	θ_β (Deg.)
18367	22.3	−22.1	90.0	44.4	67.9
18508	21.7	−22.8	90.0	44.5	67.2
18369	22.9	−21.6	−89.2	44.5	67.6
18509	67.9	−67.6	0.1	44.5	67.8

Fig. 5.21 Operability map for H1RG-18509 at a temperature of 28 K and applied bias of 150 mV, where inoperable pixels are shown in black. The (log magnitude) FFT of the operability map is shown in the upper-left corner. The FFT image was scaled to show a better contrast between the line features and the background

other dark current mechanisms have several other parameters to fit, which can lead to a non-unique set of fitted parameters. Figure 5.24a shows an example of the dark current as a function of bias for an operable pixel in H1RG-18508 at large applied bias, where the behavior of the pixel closely matches the theory of band-to-band tunneling. Inoperable pixels required the addition of trap-to-band tunneling to fit the I-V data. Figure 5.24b shows an inoperable pixel in H1RG-18508, where the effective-β parameter of the operable nearest neighbor pixels was used to fit band-to-band tunneling current, and is shown to be orders of magnitude below the measured dark current and hardly contributes to the fit.

Unlike band-to-band tunneling where β is the only unknown parameter, trap-to-band tunneling requires the fitting of five additional parameters (see Sect. 3.3). Furthermore, the fitted trap-to-band tunneling heavily depends on the parameters' initial guesses, which may not be unique.

Therefore, only band-to-band is initially fitted to the larger applied bias I-V data of operable pixels to estimate the value of β. The following step to model the dark current was to then fit the thermal dark currents (diffusion and G-R) to the higher temperature I-T data. Lastly, the test dewar light leak or "mux glow" is fitted, along with trap-to-band tunneling if necessary.

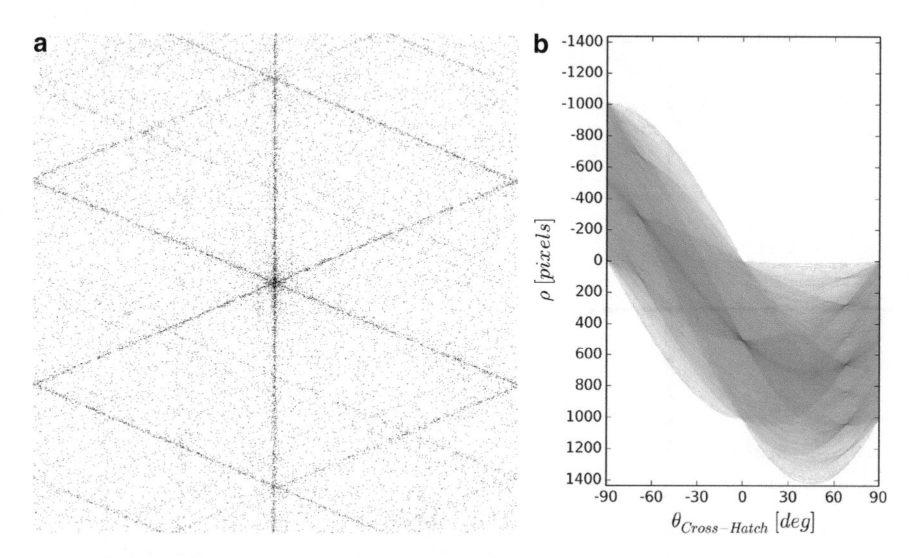

Fig. 5.22 Binarized input image (**a**) and Hough transform (**b**) to determine the cross-hatching angles in H1RG-18509. The input image is the FFT of the operability map for H1RG-18509 at a temperature of 28 K and an applied bias of 150 mV. A threshold of 1.5 standard deviations above the mean was used to binarize the image shown on the top-left corner of Fig. 5.21

When estimating the value of β in H1RG-18367 and H1RG-18509, an additional constant current was added to the I-V data along with band-to-band tunneling (similar to that shown on Fig. 5.27 on individual pixels) because the large bias data was not completely dominated by band-to-band tunneling. In the case of H1RG-18367, the "glow" from the multiplexer (discussed in Sect. 5.1.2 and throughout 5.1.4.1) has a considerable effect on the I-V data up to 300 mV of bias, while H1RG-18509 is dominated by a possible $<1\ e^-/s$ light leak in the test dewar and G-R up to a bias of $\sim$350 mV. Since G-R current does not have a large bias dependence in the range of interest, the light leak plus the G-R current in H1RG-18509 were modeled as a constant current when estimating β.

Figure 5.25 shows the distribution of fitted β values to a 32$\times$400 pixel region in the four arrays presented here. The fit of dark current models to the data is limited to the subarray used to obtain warmup data (see Sect. 4.7). The central 400 columns were used to avoid any significant changes in cutoff wavelength amongst the pixels used to show the uniformity of band-to-band tunneling current.

5.2.2 Dark Current Model Results

Figure 5.26 shows the dark current Arrhenius plot at a detector bias of 276 mV and dark current *vs.* bias (at a temperature of 28 K) for 36 operable pixels in H1RG-18508. The I-T data shown is the average of all 36 pixels, where the error

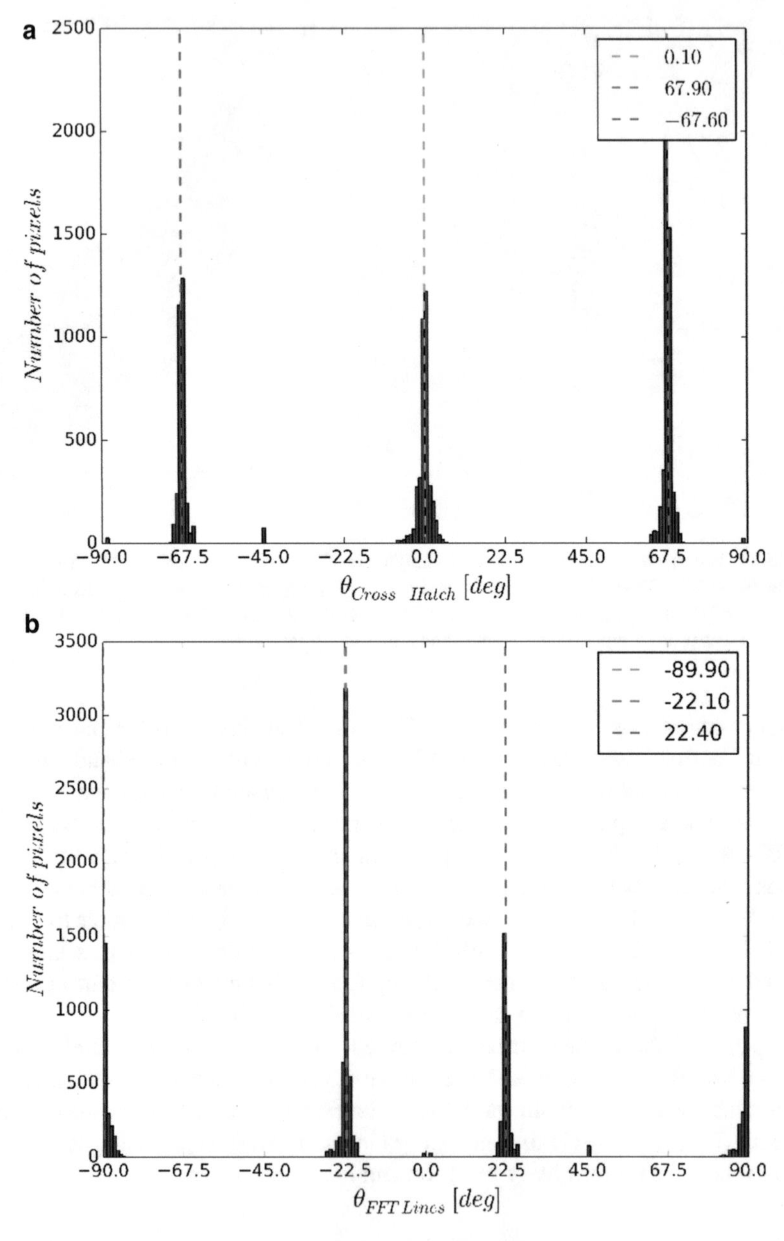

Fig. 5.23 Angle distribution with respect to the horizontal of the cross-hatching pattern in the operability map of H1RG-18509 (**a**) and that corresponding in Fourier space (**b**). The cross-hatching angle (**a**) distribution is obtained from the output of the Hough transform, where this same distribution is simply rotated by 90° to obtain that of the line features in the FFT image of the operability map (used as the input image to the Hough transform)

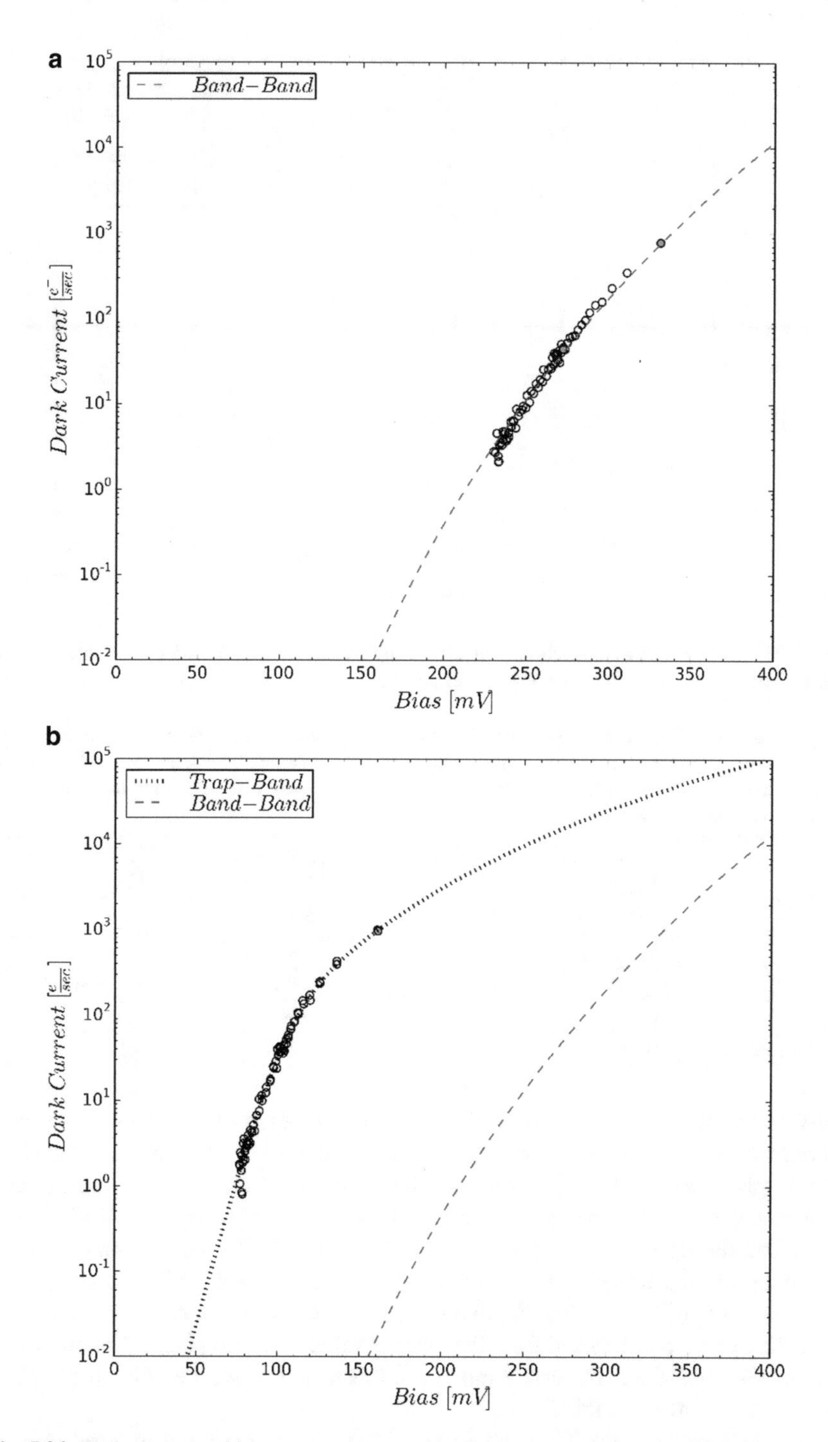

Fig. 5.24 Dark current *vs.* bias at a temperature of 28 K for (**a**) an operable and (**b**) inoperable (at 150 mV of applied bias) pixel in H1RG-18508

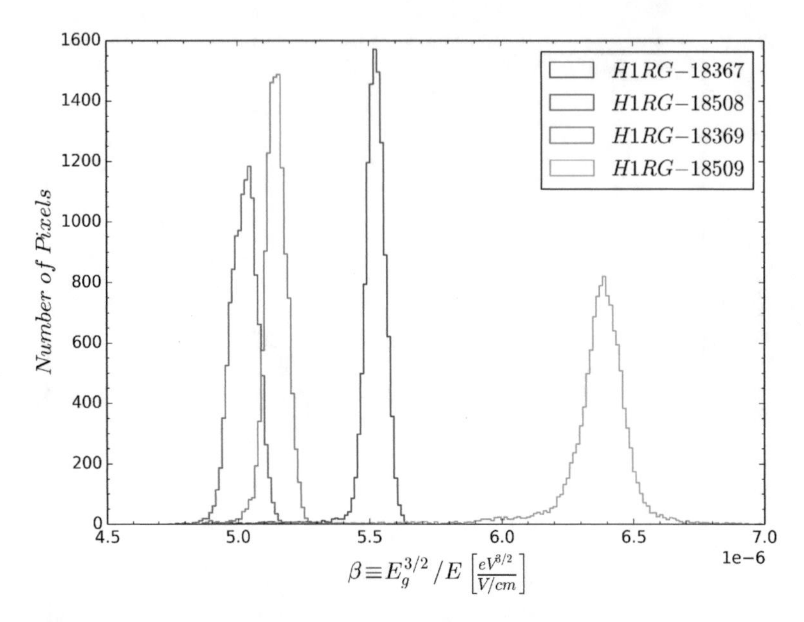

Fig. 5.25 Fitted β value distribution, with an actual reverse bias of 300 mV, to operable pixels in the central 32 rows (for which we have warm-up data) and 400 columns of the array

Table 5.12 Optimized dark current parameters to model the mean dark current behavior of a collection of pixels for all four LW13 arrays shown in Figs. 5.26, 5.27, 5.28, and 5.29. Note that the "Light Leak" parameter for H1RG-18367 corresponds to the glow from the multiplexer, and not an actual light leak in the test dewar

	H1RG-18508 Fig. 5.26	H1RG-18367 Fig. 5.27	H1RG-18369 Fig. 5.28	H1RG-18509 Fig. 5.29
$\beta\ \left[\frac{eV^{3/2}}{V/cm}\right]$	5.03×10^{-6}	5.53×10^{-6}	5.15×10^{-6}	6.39×10^{-6}
$\tau_h\ [s]$	1.4×10^{-7}	–	1.8×10^{-7}	3.7×10^{-7}
$\tau_{GR}\ [s]$	1.9×10^{-5}	–	9.8×10^{-6}	8.9×10^{-6}
$E_{t_{gr}}\ [eV]$	0.062	–	0.067	0.067
Light Leak $\left[e^-/s\right]$	0.09	8.8	0.28	0.19

bars on the squares correspond to the standard deviation of the mean in the dark current values for the averaged pixels. The individual I-V data curves for each of the 36 pixels are displayed along with the mean initial dark current and detector bias (squares) for the 36 pixels corresponding to 150, 250, and 350 mV of applied bias. The shaded region's upper and lower bounds coincide with the band-to-band tunneling calculated from the β values that are two standard deviations away from the mean of the fitted β value distribution in Fig. 5.25, while the single band-to-band tunneling curve is calculated from the mean β value of the 36 pixels. The optimized parameters to fit the dark current models to this data, and for the other three LW13 devices are shown in Table 5.12.

The I–V data for the 36 pixels in Fig. 5.26b follow the trend of band-to-band tunneling with slightly different β values at biases >200 mV. The dark current models fitted to the I-T data shows that, up to a temperature of ~32 K, the

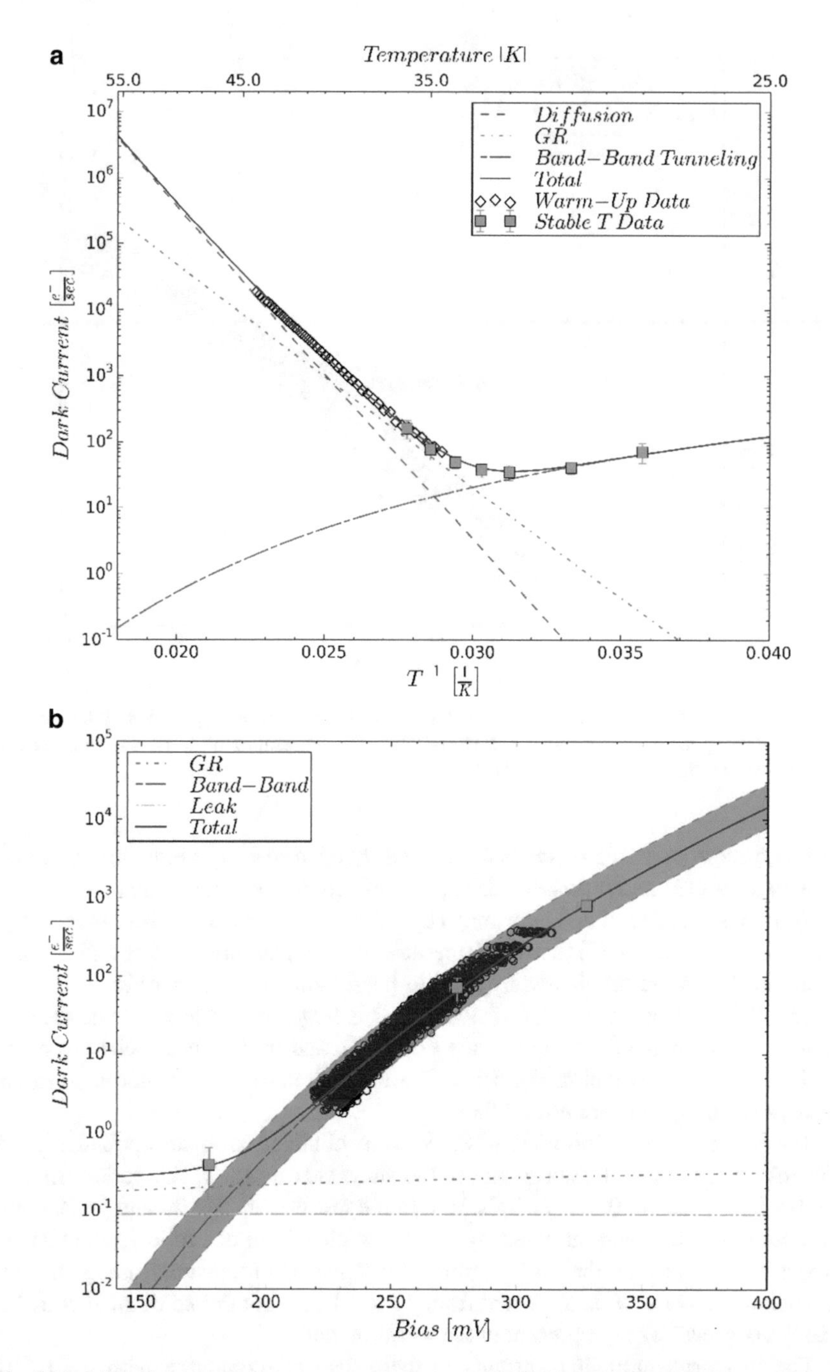

Fig. 5.26 (**a**) Dark current *vs.* temperature (mean actual back-bias of 276 mV) averaged data of 36 operable pixels in H1RG-18508. The error bars correspond to ± one standard deviation of the mean in the dark current values for the averaged pixels. (**b**) Individual dark current *vs.* bias (at 28 K) data curves (empty circle data points) for the same 36 pixels shown in (**a**). The cyan square data points are the average of the initial dark current and detector bias for the 36 pixels. The shaded region in (**b**) corresponds to the band-to-band tunneling calculated from the β value ± two standard deviations from the mean, while the single band-to-band tunneling curve corresponds to that of the mean β value of the 36 pixels

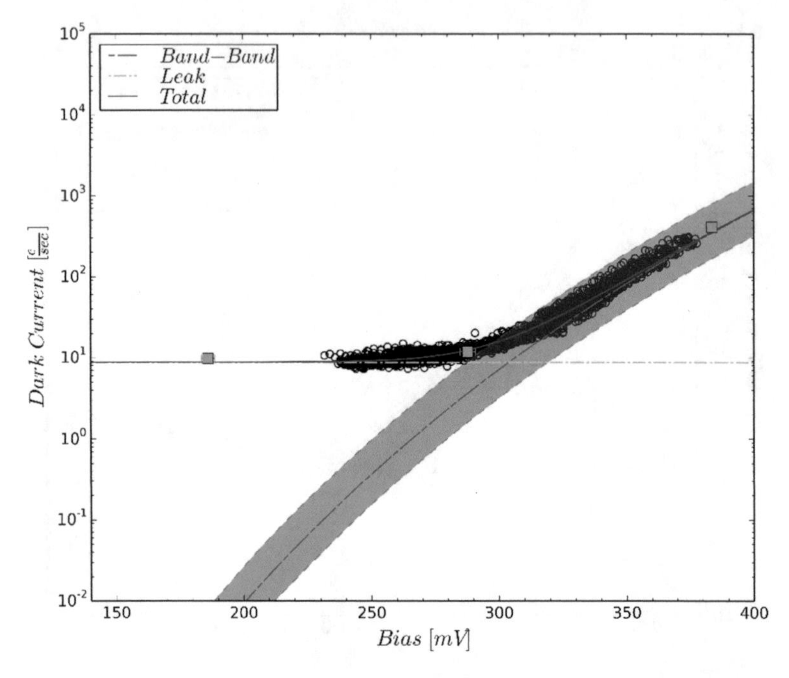

Fig. 5.27 Dark current *vs.* bias data curves for 36 pixels which where operable at a temperature of 28 K and applied bias of 250 mV in H1RG-18367 at a temperature of 28 K. The constant leak current that was fitted corresponds to the "mux glow"

dark current of operable pixels is dominated by band-to-band tunneling. At higher temperatures G-R and diffusion currents are the dominating components.

Similar to H1RG-18508, the majority of the pixels in the other three arrays (Fig. 5.27, 5.28, and 5.29) that were operable at a temperature of 28 K and 250 mV of applied bias were dominated by band-to-band tunneling at larger biases.

The observed "mux glow" in H1RG-18367 is large enough to dominate the dark current up to a bias of ~300 mV (see Fig. 5.27), and in the linear behavior of the SUTR curve for the pixel in Fig. 5.10. Thermal dark currents were not fitted to this array as warm-up data are not available.

The warm-up data for H1RG-18369 were obtained with an applied bias of 150 mV, where the I-T data in Fig. 5.28a show that at 28 K, the dark current is limited by a possible 0.3 e^{-}/s light leak in the test dewar. The 34 and 35 K stable data points, which were affected by the "mux glow", do not follow the behavior expected from any of the dark current mechanisms, further confirming that the anomalous increase of dark current from 33 to 34 K is not due to thermal currents. The "mux glow" was not present in the warm-up data.

The increased tunneling currents in these devices, compared with the LW10 devices, was expected given the smaller band gap. H1RG-18509 was designed to address this concern. A substantial decrease in band-to-band tunneling current in this array compared with the other three arrays can be seen in Fig. 5.29. The small

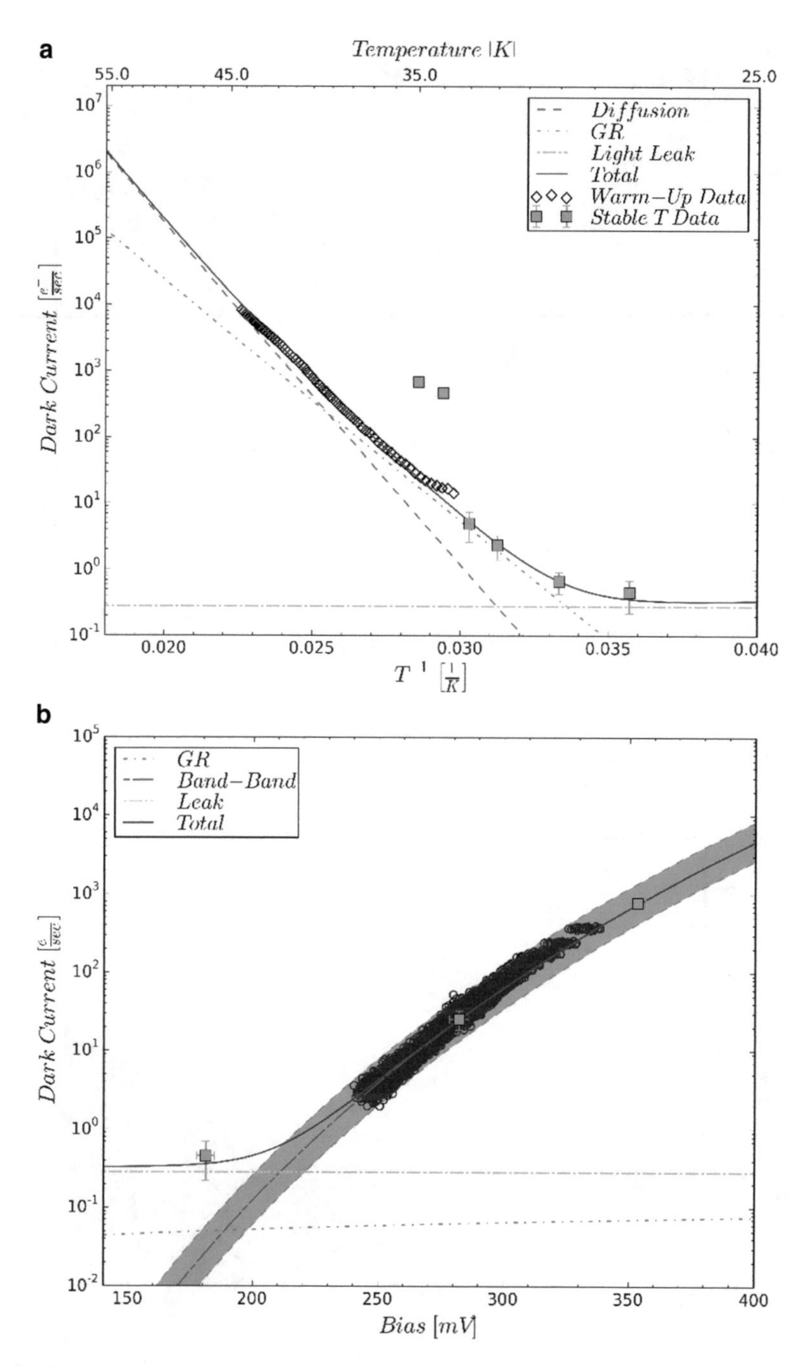

Fig. 5.28 (**a**) Dark current *vs.* temperature (applied bias of 150 mV) averaged data of 36 operable pixels in H1RG-18369. The mean detector bias of 181 mV, corresponding to the 28 K stable data point, was used to fit the dark current models. (**b**) Individual dark current *vs.* bias (at 28 K) data curves for the same 36 pixels shown in (**a**). The 33 and 34 K stable temperature data points in (**a**) were affected by a "mux glow", increasing the dark current above the expected value from thermal dark currents

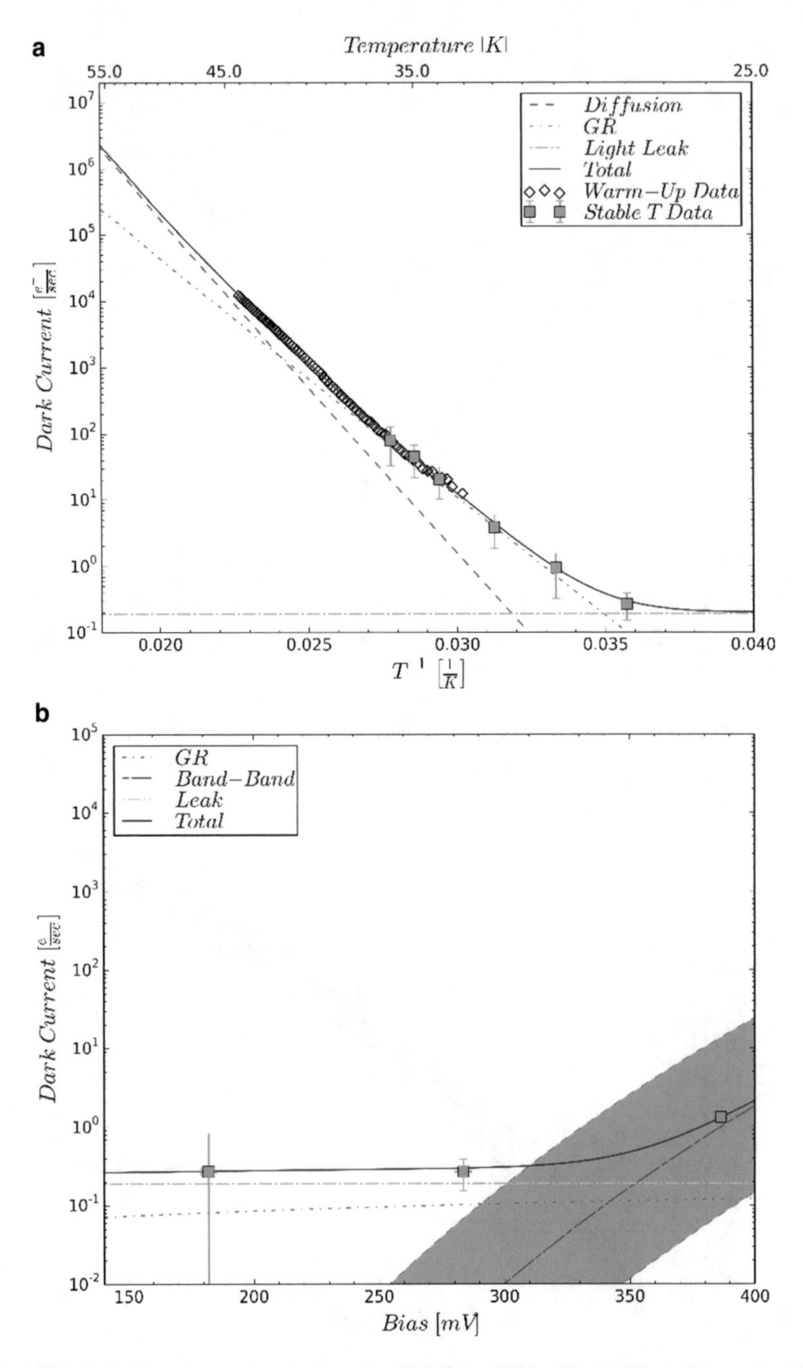

Fig. 5.29 (**a**) Dark current *vs.* temperature (applied bias of 250 mV) and (**b**) dark current *vs.* bias (at 28 K) averaged data of 50 operable pixels in H1RG-18509. In (**a**), the mean detector bias of 284 mV, corresponding to the 28 K stable data point, was used to fit the dark current models

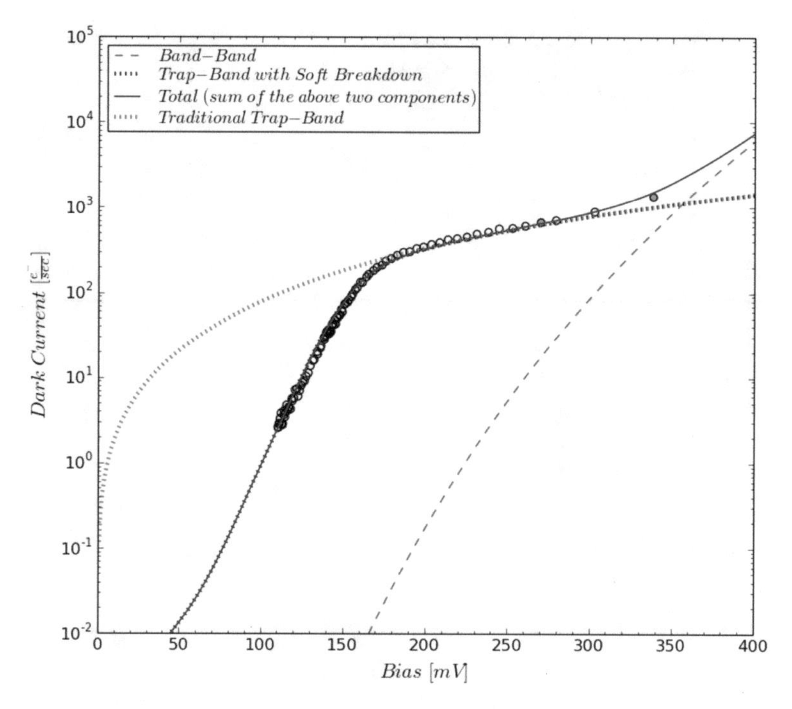

Fig. 5.30 Dark current *vs.* bias at a temperature of 28 K for inoperable pixel [534, 631] in H1RG-18369. Trap-to-band tunneling with a constant trap density and with variable trap density (due to activated traps) are plotted for comparison. The total curve corresponds to the sum of band-to-band and trap-to-band with soft breakdown curves

curvature in the SUTR curves (shown in the double derivative at the beginning of the SUTR curves in Fig. 5.15) for the majority of the pixels for this array is due to the relatively small tunneling current found in the I-V data (Fig. 5.29b). As a consequence, we are able to achieve larger well depths by applying a large bias ($\sim$75 ke^- with an applied bias of 350 mV). The warm-up data for this array were obtained with an applied bias of 250 mV. Dark current data for this array above 29 K is dominated by G-R and diffusion, while lower temperature dark current which is well below 1 e^-/s approaches the light leak level.

The apparent dominance of band-to-band tunneling at higher biases in operable pixels for all arrays is very encouraging, as further enhancements in Teledyne's design to increase β will decrease this tunneling component in future longer wavelength devices.

5.2.2.1 Modified Trap-to-Band Tunneling Fit

Figure 5.30 shows the I-V data of an inoperable pixel which can be modeled with trap-to-band tunneling following a soft breakdown. Below 170 mV of applied

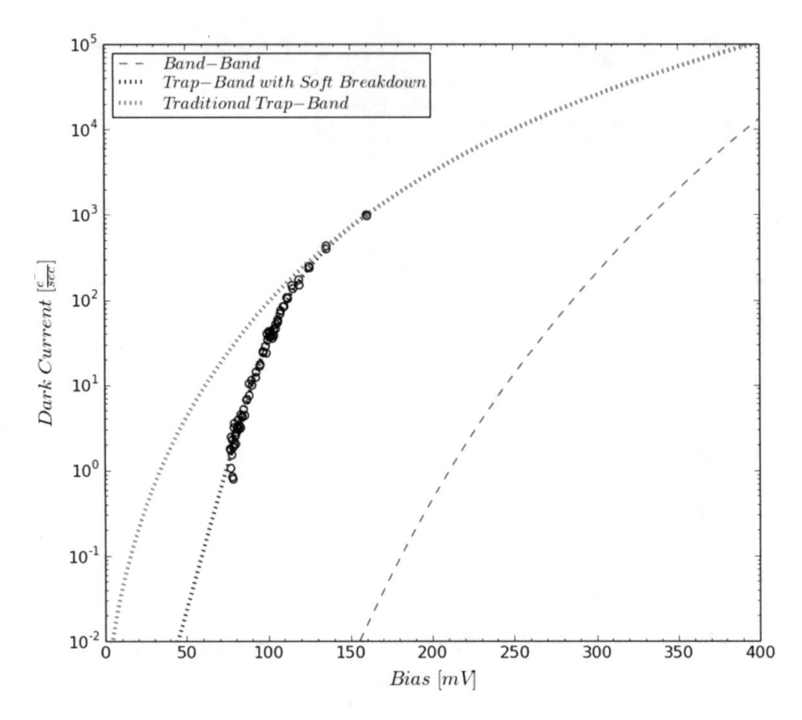

Fig. 5.31 Dark current *vs.* bias at a temperature of 28 K for inoperable pixel [514, 450] in H1RG-18508 shown in Fig. 5.24, where the trap-to-band tunneling with a constant trap density and with variable trap density (due to activated traps) are plotted for comparison. The total curve corresponds to the sum of band-to-band and trap-to-band with soft breakdown curves

reverse bias, trap-to-band tunneling with the assumption of a constant trap density at all biases (labeled as "Traditional Trap-Band") does not fit the data, requiring the use of Eq. 3.26 which allows for the activation of traps. Similar behavior can be seen among inoperable pixels in all four LW13 arrays, where the data following the soft breakdown can be modeled by trap-to-band tunneling. Figure 5.31 shows the inoperable pixel presented in Sect. 5.2.1 where we show the comparison between the trap-to-band expected from a constant trap density and the activation of traps. This unusual behavior was seen in the dark current data on LW diodes from TIS presented by Bailey et al. [18] and by Bacon et al. [21]. The fitted parameters to model the data of these two inoperable pixels are in Table 5.13.

More data at lower biases would be needed to more accurately characterize the threshold voltage of the traps which contribute to the trap-to-band tunneling, while data at larger biases is needed to properly characterize pixels which only appear to show the soft break down, and there is no indication if the behavior before or after the soft breakdown is consistent with trap-to-band tunneling or any other form of dark current.

Table 5.13 Optimized dark current parameters to model the dark current behavior of inoperable pixels [534, 631] in H1RG-18369 and [514, 540] in H1RG-18508 with the traditional and modified trap-to-band tunneling currents. The β parameter used for both figures corresponds to the mean of their corresponding operable four nearest neighbors

	Figure 5.30	Figure 5.31
$\beta\ \left[\frac{eV^{3/2}}{V/cm}\right]$	5.11×10^{-6}	5.05×10^{-6}
$E_t\ [eV]$	0.08	0.05
$n_{t_i}\ \left[cm^{-3}\right]$	1	3.3×10^5
$n_{t_d}\ \left[cm^{-3}\right]$	2.2×10^3	4.4×10^8
$V_a\ [mV]$	154	107
γ	0.2	0.23

5.3 Discussion on Constant Dark Current Regimes

We have shown that tunneling dark currents dominate the operability of these devices, where the bias-dependent effects of these currents may present a problem in the calibration of low signal data.

The reduction of tunneling dark currents in H1RG-18509 allows for a constant dark current calibration at low signals since the dark charge *vs.* time is nearly linear at biases as large as 350 mV. The non-linear effects on collected signal *vs.* time due to large band-to-band tunneling currents seen in the other three arrays at large biases may be modeled for each of the operable pixels in an attempt to calibrate it, but the exact (or very close to) initial detector bias across each pixel is needed since the tunneling currents vary appreciably as pixels debias.

Alternatively, the three arrays that show large band-to-band currents can be operated in regimes where the dark current is not dominated by band-to-band tunneling.

In the low applied bias regime where band-to-band tunneling is negligible, thermal dark currents would be the dominant source of dark current and could be calibrated at stable temperatures since these currents are approximately constant (above ~50 mV of reverse bias) as pixels debias. For applications that require larger well depths than can be attained with the Hawaii-XRG multiplexers at the biases needed to operate in this regime (~<200 mV), a capacitive transimpedance amplifier (CTIA) multiplexer could be used in its place. A CTIA multiplexer would allow the operation of these arrays at a constant small bias voltage while signal is integrated, therefore maintaining a constant dark current, and with a much larger well depth. Alternatively, capacitance could be added to the integrating node, as was done for the 512×512 arrays analyzed by Bacon [20].

At higher temperatures, the thermal dark current mechanisms, will dominate. Tunneling currents decrease with increasing temperature since the band gap energy increases. At higher temperatures, the dark current due to thermal currents may be comparable to those from band-to-band tunneling at lower temperatures. Figure 5.26a shows that with an applied bias of 250 mV, the dark current in H1RG-

18508 at 35 K is dominated by G-R and is comparable to that at 28 K which is dominated by band-to-band tunneling.

5.4 Phase I Summary

All four LW13 detector arrays had cutoff wavelengths ranging from 12.4 to 12.8 μm, measured at a temperature of 30 K. Two of the arrays (H1RG-18367 and 18508) had the same design as the LW10 detector arrays designed for the proposed NEOCam mission[24–26] extrapolated to longer wavelengths, while the other two arrays (H1RG-18369 and 18509) had two different TIS proprietary designs with the goal of mitigating quantum tunneling dark currents.

The median dark current per pixel for three of the four arrays was below 1 e^-/s at a temperature of 28 K and 150 mV of applied reverse bias, with a median well depth of $\sim$43 ke^-. We were able to show that the dark current is dominated by G-R[28] and diffusion[12] as the operating temperature is increased, while increasing the bias (for larger well depth) increases quantum tunneling dark currents exponentially.

LW13 array H1RG-18509, designed to mitigate the effects of quantum tunneling dark currents, had the best dark current and well depth performance at larger applied reverse biases. At a temperature of 28 K and applied bias of 350 mV, 86% of the pixels had dark currents below 10 e^-/s and well depth of at least 75 ke^- (median dark current and well depth of 1.8 e^-/s and 81 ke^- respectively). The other three LW13 arrays H1RG-18367, 18369, and 18508 had median dark currents of about 379, 730, and 780 e^-/s at 28 K and applied bias of 350 mV, where the dark current at this bias is dominated by band-to-band or trap-to-band (defect assisted) tunneling.

An increase in band-to-band and defect assisted tunneling current was expected array-wide as the cutoff wavelength is increased to a target of 15 μm due to the smaller band-gap. The effect of trap-to-band is less predictable than band-to-band tunneling since it is dependent on the defect/dislocation density of individual pixels.

The characterization results from the LW13 arrays suggests that a similar array design used for H1RG-18509 would be the best approach for the final phase of the project to extend the cutoff wavelength to 15 μm to reduce quantum tunneling dark currents.

Chapter 6
Phase II Results: 15 μm Cutoff Wavelength Devices

The UR infrared detector group received three 1024 × 1024 pixel detector arrays (H1RG-20302, H1RG-20303, and H1RG-20304) bonded to H1RG multiplexers for the final phase of the project from TIS. Table 6.1 includes the quantum efficiency (QE) and cutoff wavelength measurements provided by TIS for the three LW15 arrays at a temperature of 30 K from the PECs that were grown and processed at the same time as the megapixel arrays. The characterization of these arrays is carried out in the same manner as the LW13 arrays, only at lower temperatures to decrease dark currents due to the reduced band gap energy in these devices with respect to the LW13 devices. The results of dark current model fits and characterization (calibration, CDS read noise, dark current, and well depth) are presented here. Unlike the LW13 arrays, operability here is used only to show that a subset of large dark current/low well depth pixels in these arrays also form the same cross-hatching pattern as the LW13 arrays. Since the cutoff wavelength of one of these devices is considerably longer than the other two, similar dark current and well depth requirements would not provide an adequate quantitative performance comparison between the arrays. The main focus of this chapter will be to show the dark current behavior as a function of temperature and bias of these arrays.

6.1 Characterization

In addition to characterizing these arrays at lower temperatures than the LW13 arrays, data are also obtained at one additional lower bias. Capacitance, dark current, well depth, and non-linearity have all been measured at biases of 50, 150, 250, and 350 mV.

M. Cabrera, *Development of 15 Micron Cutoff Wavelength HgCdTe Detector Arrays for Astronomy*, Springer Theses, https://doi.org/10.1007/978-3-030-54241-2_6

Table 6.1 Cutoff wavelength and QE measurements for all three LW15 arrays were provided by TIS at a temperature of 30 K with an applied reverse bias of 100 mV. These arrays do not have anti-reflective coating. The PEC QE measurements without anti-reflective coating are above the theoretical value (78%), but are within experimental measurement error

Detector H1RG-	Wafer	Cutoff wavelength (μm)	QE (6–12 μm)
20302	3995	16.7	81%
20303	3994	15.5	83%
20304	4018	15.2	80%

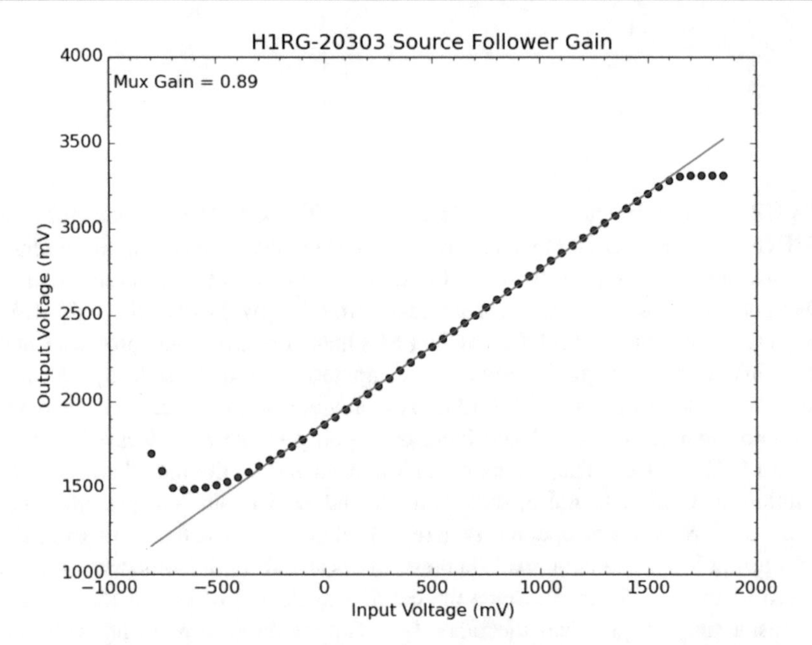

Fig. 6.1 Output *vs.* input referred signal curve for H1RG-20303 used to calculate the gain from the multiplexer at a temperature of 25 K. The multiplexer gain corresponds to the slope in the linear region of the curve. The voltage at the integrating node for these devices will vary from the reset voltage of 100 mV (in addition to any charge injection) to approximately the substrate voltage if the device is saturated (typically as large as 450 mV, corresponding to an applied bias of 350 mV).

6.1.1 Calibration

The source-follower FET gain in the multiplexer to convert between the output and input-referred signal was 0.9 for all three LW15 arrays, shown for H1RG-20303 in Fig. 6.1.The nodal capacitance was measured at the above mentioned biases and at a temperature of 25 K for H1RG-20303 and 20304, and at 23 K for H1RG-20302 because of the slightly longer cutoff wavelength in order to reduce the dark current.

Figure 6.2 shows the distribution of nodal capacitance per pixel at each of the four applied biases. The apparent spread in capacitance is mostly due to uncertainties in this method. These capacitance values are not yet corrected for IPC. The IPC

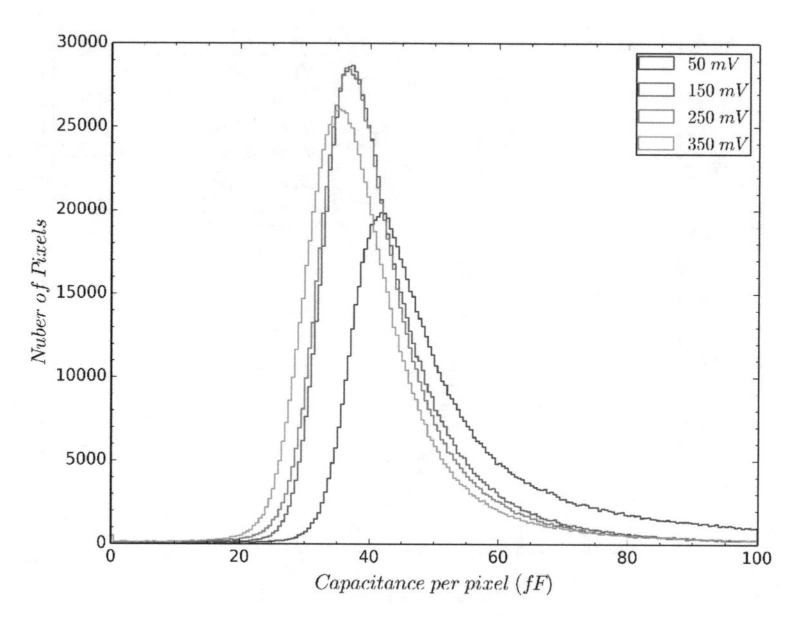

Fig. 6.2 Capacitance per pixel distribution for H1RG-20303 at a temperature of 25 K and at all tested applied biases. These capacitances have not yet been corrected for interpixel capacitance. The spread in the capacitance distribution is primarily due to uncertainty in our probabilistic determination

Table 6.2 The multiplexer (source-follower FET) gain, IPC coupling parameter α, and the median IPC corrected capacitance in femtofarads for each of the applied biases is given for each of the LW15 arrays. All measurements in this table for H1RG-20303 and 20304 were made at a temperature of 25 K, and at a temperature of 23 K for H1RG-20302

			Median capacitance (fF) corrected for IPC			
Detector H1RG-	Mux gain	α (%)	50 mV	150 mV	250 mV	350 mV
20302	0.9	0.82	43	34	36	35
20303	0.9	0.88	44	37	36	35
20304	0.9	0.76	49	44	39	36

corrected nodal capacitance, source-follower FET gain, and α parameter value for all three arrays are shown in Table 6.2. The corrected nodal capacitance is used to determine the conversion between measured voltages and electrons.

6.1.2 CDS Read Noise

The CDS read noise was measured at a bias of 50 mV and temperatures of 23 K for H1RG-20302 and 20304, and at 24 K for H1RG-20303. Figure 6.3 shows the rms CDS total noise per pixel for all three arrays. The median CDS noise for these

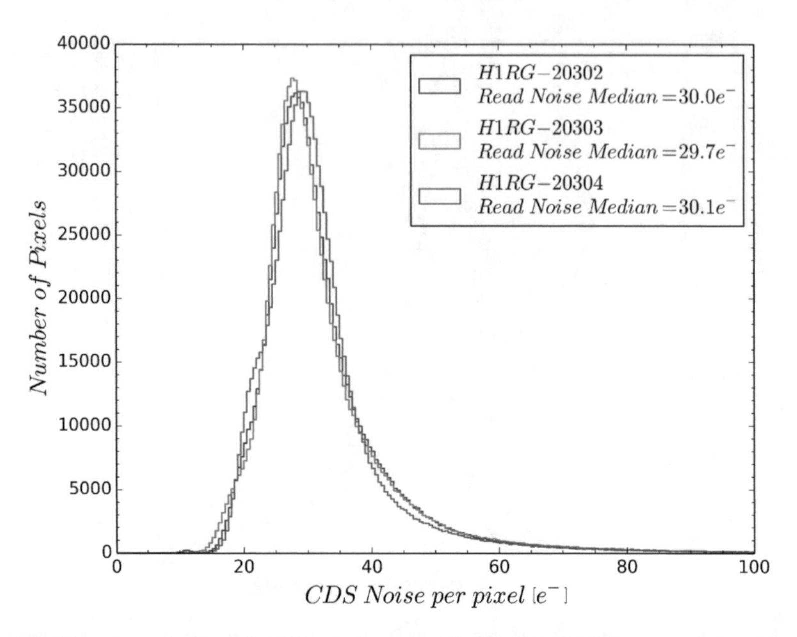

Fig. 6.3 Histogram of rms total CDS noise per pixel for the three LW15 arrays at a temperature of 23 K for H1RG-20302 and 20304, and 24 K for H1RG-20303, both with an applied reverse bias of 50 mV

devices is $\sim$7 e^- larger than that measured for the LW13 devices. If the median dark current is subtracted in quadrature as "shot-like" noise (i.e. $\sigma^2_{dark\ current} = I_{dark} \cdot t_{integration\ time}$), the median CDS noise is reduced to 27.6, 27.1, and 29.9 e^- for H1RG-20302, 20303, and 20304 respectively. The increased CDS read noise is most likely due to the quality of the multiplexer itself since the HAWAII muxes used in this project are engineering grade, or to the low operating temperature. Carrier freeze out in silicon CMOS devices becomes an issue at low temperatures, where the number of available carriers from ionized impurity atoms that contribute to conduction decreases exponentially with temperature (Eq. 1.9). Due to the ionization probability being less than unity at these low temperatures, ionization and recombination of carriers will affect the conductivity of the multiplexer and appear as noise [48]. The HAWAII muxes were designed to operate at temperatures between 30–300 K [49], and the lower operating temperature could have contributed to the small increase in read noise.

6.1.3 Dark Current, Well Depth, and Cross-Hatching

Tables 6.3, 6.4, and 6.5 show the median dark current and well depth of the three LW15 devices measured at temperatures between 23–30 K and applied biases of

50, 150, 250, and 350 mV of applied reverse bias. Since an increase in tunneling currents was anticipated due to the increased cutoff wavelength, dark current and well depth were measured at the lower applied bias of 50 mV. Furthermore, lower temperature data was obtained to reduce thermal currents.

Table 6.3 Median dark current and well depth with different applied biases and temperatures for H1RG-20302

	Median dark current (e^-/s)			
	Median well depth (ke^-, mV)			
Temperature	Bias = 50 mV	150 mV	250 mV	350 mV
23 K	11	32	635	592
	20, 76	38, 178	55, 245	53, 243
24 K	22	49	653	644
	20, 75	37, 178	55, 247	53, 245
25 K	54	88	695	730
	20, 74	38, 176	56, 249	54, 248
26 K	140	163	780	870
	19, 71	37, 173	55, 248	54, 248
27 K	307	333	941	1110
	18, 67	36, 168	55, 244	54, 247
28 K	565	673	1284	1546
	16, 61	34, 160	53, 236	53, 241

Table 6.4 Median dark current and well depth with different applied biases and temperatures for H1RG-20303

	Median dark current (e^-/s)			
	Median well depth (ke^-, mV)			
Temperature	Bias = 50 mV	150 mV	250 mV	350 mV
23 K	7	16	263	790
	20, 75	41, 176	62, 274	66, 306
24 K	13	29	272	804
	20, 74	40, 175	62, 273	67, 308
25 K	29	53	301	858
	20, 73	40, 174	62, 272	67, 311
26 K	69	104	373	943
	20, 72	40, 173	61, 270	68, 314
27 K	179	205	478	1080
	19, 69	39, 169	61, 267	68, 313
28 K	350	427	649	1307
	18, 65	38, 165	59, 261	67, 310
29 K	646	837	1107	1712
	16, 60	37, 159	57, 253	66, 305
30 K	1275	1848	2135	2606
	14, 53	35, 150	55, 240	63, 292

Table 6.5 Median dark current and well depth with different applied biases and temperatures for H1RG-20304

	Median dark current (e^-/s)			
	Median well depth (ke^-, mV)			
Temperature	Bias = 50 mV	150 mV	250 mV	350 mV
23 K	0.8	1.4	30	649
	25, 83	49, 182	68, 282	78, 345
24 K	1.1	2	27	638
	25, 80	49, 180	68, 280	78, 344
25K	2.4	3.6	24	620
	24, 78	48, 178	67, 278	79, 346
26 K	6	8	24	602
	23, 77	48, 176	67, 276	78, 346
27 K	15	19	32	585
	23, 76	48, 176	67, 276	80, 354
28 K	38	47	58	582
	23, 76	48, 175	67, 276	81, 356
29 K	94	112	119	604
	23, 75	47, 175	67, 275	81, 358
30 K	215	257	257	672
	23, 74	47, 173	66, 273	81, 357

The improvement in the diode design from the best LW13 device H1RG-18509 is most obvious in H1RG-20304, where at a temperature of 28 K and actual reverse bias of 276 mV the median initial dark current is 58 e^-/s. This dark current is almost identical to that of H1RG-18508 (LW13 array with the standard NEOCam array design) with a median initial dark current of 57 e^-/s and an actual reverse bias of 275 mV at the same temperature of 28 K (see Table 5.6). If H1RG-18508 had a cutoff wavelength of 15.2 μm (that of LW15 array H1RG-2304), the dark current due to band-to-band alone at 28 K with a detector bias of 276 mV would be $\sim 2.5 \times 10^6 e^-/s$, nearly five orders of magnitude larger than was measured for H1RG-20304. This improvement in dark current is due to the improved array design (larger β value) of the LW15 device. Although 275 mV of detector bias was not reached for H1RG-20302 with the temperature and bias combinations shown in Table 6.3 due to larger dark currents debiasing the array before the pedestal, it will be shown in Sect. 6.2 that near 275 mV of applied bias the dark current for well behaved pixels (those that are considered operable by similar NEOCam criteria at a bias of 150 mV) is below $10^4 e^-/s$ at a temperature of 23 K. This reduction in band-to-band tunneling is remarkable given that H1RG-20302 has the longest cutoff wavelength of 16.7 μm, and the band-to-band tunneling measured for this device would decrease if measured at a temperature of 28 K, where this device also shows an improvement in array design to mitigate tunneling currents.

Unlike the LW13 devices, it is not evident from the median dark current values when tunneling currents are the main source of dark current. It will be explicitly shown in Sect. 6.2 that at 350 mV band-to-band tunneling is the dominant component of dark current, yet the median dark current for H1RG-20302 and 20303 continues to increase with increasing temperature, whereas for the LW13 devices (see Sect. 5.1.4) and for H1RG-20304, the dark current decreases with increasing temperature (at large applied biases) until thermal dark currents become the dominant source of dark currents.

Figure 6.4 shows the dark SUTR curves used to determine the dark current for H1RG-20302 and 20304 with 350 mV of applied bias and temperatures from 23–26 K. The initial curvature of the SUTR curves should decrease with increasing temperature according to tunneling dark current theory, and therefore the initial slope of the SUTR curve will decrease. A decreasing initial slope on the SUTR curves is observed with increasing temperature in H1RG-20304 but not in 20302. Though tunneling currents are the dominant component of dark current in H1RG-20302 (shown as the curvature at the beginning of the SUTR curves) with 350 mV of applied bias at the temperatures shown in the figure, thermal dark currents increase sufficiently to overcome the decrease in tunneling currents with increasing temperature. It is also important to note that the well depths of H1RG-20302 and 20303 with 350 mV of applied bias are quite low, the worst case in H1RG-20302 where the median well depth is just over 100 mV below the applied bias. The lower well depth is expected from all of the LW15 devices due to the increased cutoff wavelength, and therefore larger tunneling dark currents that debias pixels appreciably between reset and the pedestal frame. Since H1RG-20303 and 20304 have similar cutoff wavelengths, the larger well depth in H1RG-20304 can be attributed to an improvement in the pixel design over the best LW13 device (H1RG-18509).

The cross-hatching pattern is also observed among pixels with large dark currents and/or low well depth. Figure 6.5 shows the dark current *vs.* well depth distribution per pixel at a temperature of 23 K and an applied bias of 50 mV for H1RG-20303. Though operability is not used to rank the performance of the LW15 arrays, this distribution is used to identify pixels that have large initial dark current and/or low well depth. The pixels below and to the right of the dashed lines are considered to be well behaved pixels since they have relatively low dark currents (below our fiducial requirement of 200 e^-/s based on the NEOCam requirements for their LW10 arrays) and sufficient well depth (arbitrarily chosen to include the majority of pixels). Low well depth and low dark currents are indicators that those pixels have very large dark currents and have debiased substantially between the reset and the pedestal frames.

Figure 6.6 shows the "operability" map of the array at 23 K and 50 mV of applied bias, where the black pixels ("inoperable") denote those that have dark currents above 200 e^-/s, well depths below $\sim$14 ke^- (low well depth is due to small detector bias), or both. The "inoperable" pixels form a cross-hatching pattern in Fig. 6.6 lying along directions parallel to the set of three cross-hatching lines that are formed by the lattice mismatch between HgCdTe and the CdZnTe substrate[29, 30]. This cross-hatching pattern was also observed in all four LW13 arrays. The cross-hatching pattern is much more prominent in the LW15 arrays compared to the LW13 arrays.

The FFT of the operability map (lower right corner in Fig. 6.6) shows the cross-hatching pattern rotated by 90° as a set of parallel lines in three distinct directions.

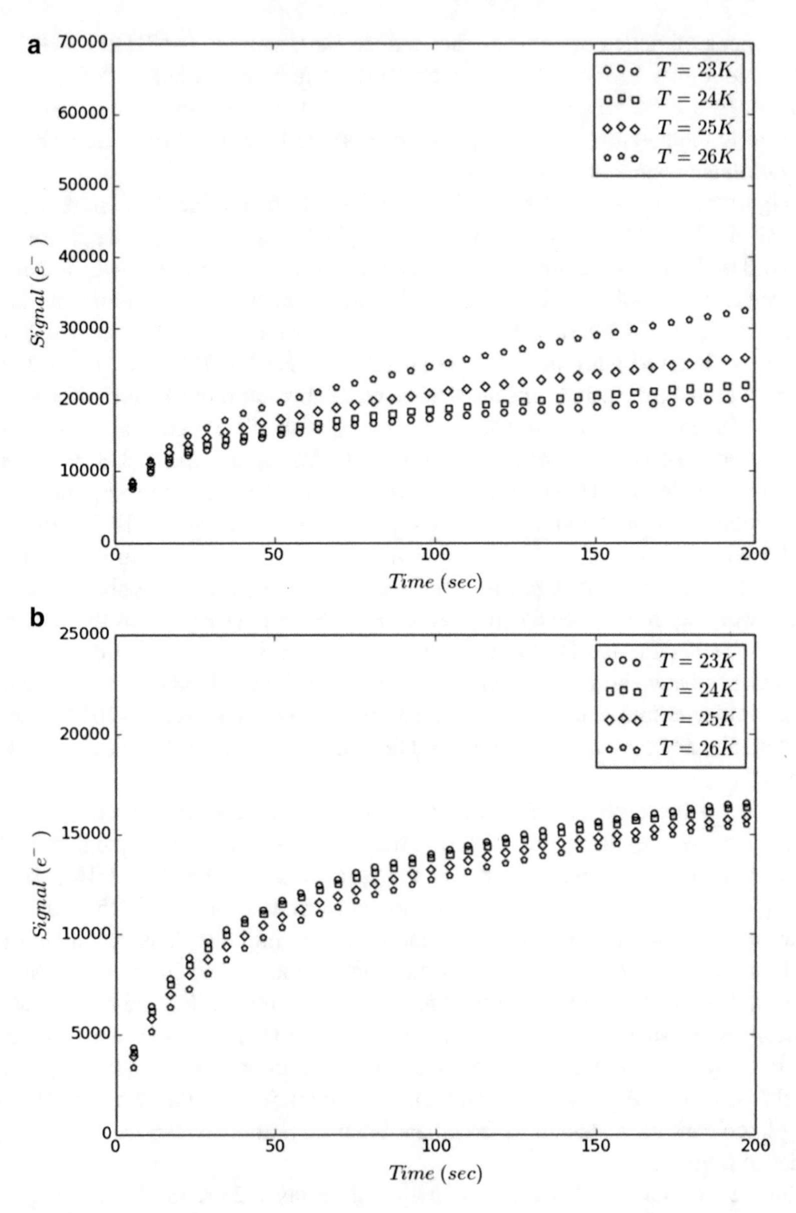

Fig. 6.4 Dark SUTR curves for (**a**) H1RG-20302 and (**b**) H1RG-20304 with an applied reverse bias of 350 mV and temperatures of 23–26 K

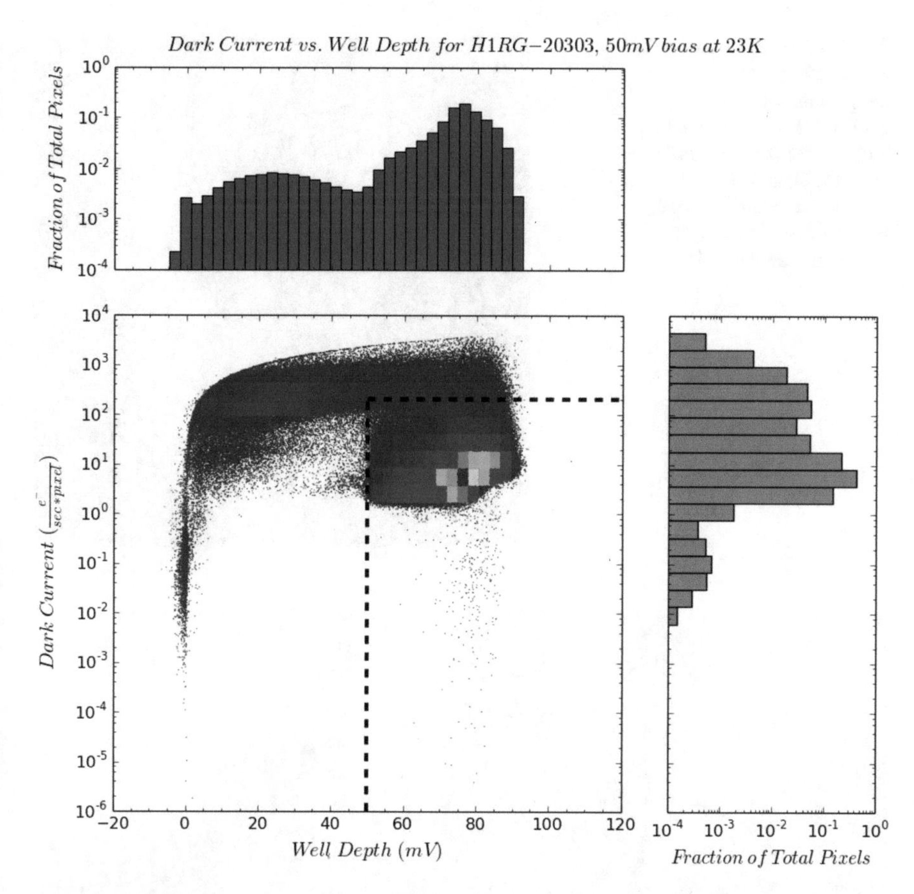

Fig. 6.5 Current in the dark *vs.* well depth distribution per pixel for H1RG-20303 at a temperature of 23 K and an applied bias of 50 mV. The vertical dashed line corresponds to well depth of ~14 ke^-. Note that the well depth corresponds to the actual initial detector bias

The prominent vertical line in the FFT corresponds to the horizontal cross-hatching pattern seen in this operability image. The large number of bad pixels at the top of the array indicates an issue with the HgCdTe growth in that region. The vertical line of bad pixels in the "operability" map may be due to an issue with the multiplexer: it corresponds to the faint horizontal line in the FFT, and is not relevant to our detector evaluation. The Hough transform was applied to the FFT of the "operability" map to estimate the angles of the cross-hatching pattern, and are consistent with those found by Martinka et al. [29] and Chang et al. [30]. The angle between the $[\bar{2}31]$ and $[\bar{2}13]$ cross-hatching lines is 44.5°, corresponding to θ_α in the cross hatching diagram shown in Fig. 4.4. The third cross-hatching pattern is parallel to the $[01\bar{1}]$ direction, and is rotated clock-wise from the horizontal axis by 0.5°. Figures 6.7 and 6.8 show the Hough transform and the angles found for the cross-hatching pattern. The other two LW15 arrays also displayed the same cross-hatching pattern as H1RG-20303 and the four LW13 arrays.

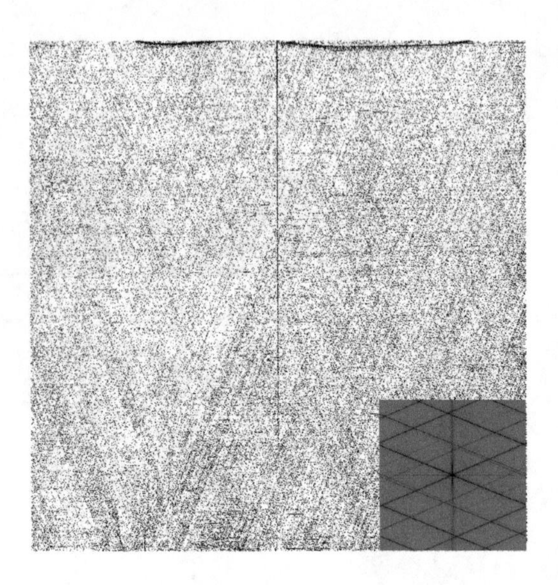

Fig. 6.6 Operability map for H1RG-20303 at a temperature of 23 K and applied bias of 50 mV, where inoperable pixels are shown in black. Operable pixels (87.3%) have dark currents below 200 e^{-}/s and well depths greater than 14 ke^{-}

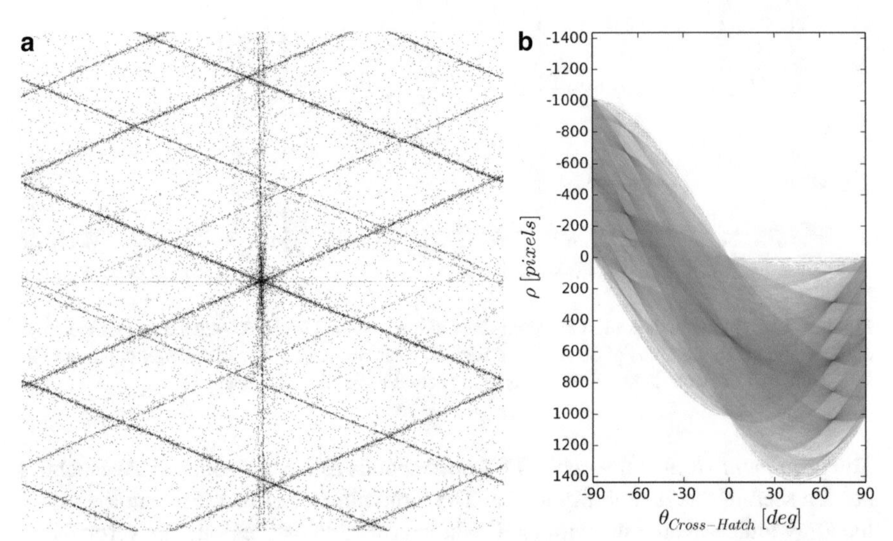

Fig. 6.7 Binarized input image (**a**) and Hough transform (**b**) to determine the cross-hatching angles in H1RG-20303. The input image is the FFT of the operability map for H1RG-20303 at a temperature of 23 K and an applied bias of 50 mV. A threshold of 1.5 standard deviations above the mean was used to binarize the image shown on the bottom-right corner of Fig. 6.6

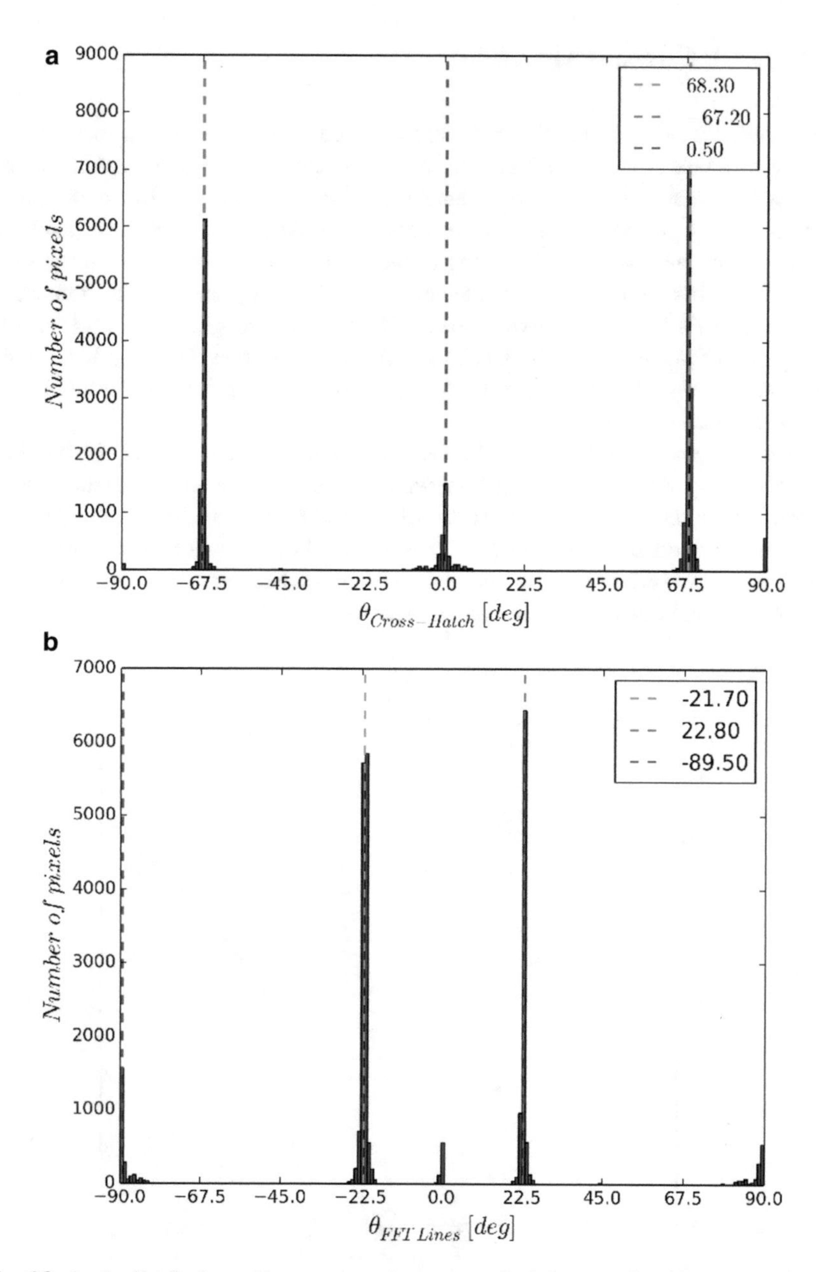

Fig. 6.8 Angle distribution with respect to the horizontal of the cross-hatching pattern in the operability map of H1RG-20303 (**a**) and that corresponding in Fourier space (**b**). The cross-hatching angle (**a**) distribution is obtained from the output of the Hough transform, where this same distribution is simply rotated by 90° to obtain that of the line features in the FFT image of the operability map (used as the input image to the Hough transform)

6.2 Dark Current Model Fits

The same method of fitting the dark current model to the LW13 data was used for these arrays (see Sect. 5.2.1). The first step in the process of fitting the dark current models is to first estimate the β parameter. Pixels with relatively low dark current and sufficient well depth were used to estimate the β parameter since these pixels are dominated by band-to-band tunneling at the largest applied biases. To distinguish these pixels from those that are affected substantially by trap-to-band tunneling, the operable *vs.* inoperable pixels labels will be used throughout this section. The operability of pixels is determined at 150 mV of applied bias and at 23 K since the warmup data was obtained at this bias, and the I-V curves at a temperature of 23 K are used to fit the tunneling currents.

Operable pixels in H1RG-20304 appear to be dominated exclusively by band-to-band tunneling at the higher biases, shown in Fig. 6.9, where any contribution from trap-to-band tunneling (if present) was not observed for this array beyond ~ 280 mV among the individual I-V curves that were studied. In cases where trap-to-band was observed, there were several points on the I-V curve that were dominated primarily by band-to-band tunneling, allowing us to get an accurate estimate of β.

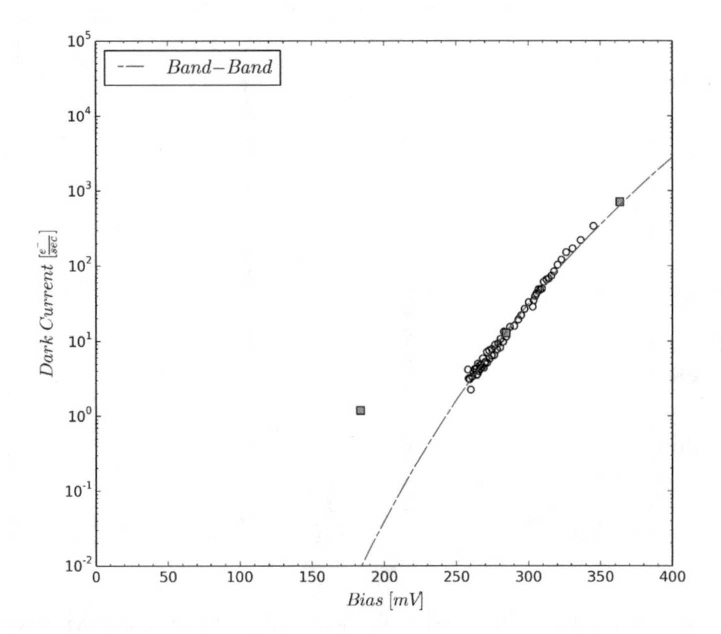

Fig. 6.9 Dark current *vs.* bias for H1RG-20304 at a temperature of 23 K. The β parameter was fitted using the I-V curve obtained from the 250 and 350 mV SUTR curves used to measure dark current and well depth. The initial dark current and well depth for this pixel with 150 mV of applied bias (~180 mV of initial detector bias) was added to show that band-to-band tunneling only affects the larger bias data. It will be shown in Fig. 6.12 that the lower bias data not affected by band-to-band tunneling is affected by a combination of light leak and trap-to-band tunneling

The other two LW15 arrays (H1RG-20302 and 20303) appeared to have a larger trap-to-band tunneling contribution at the largest applied bias of 350 mV, where only 2–3 data points at the largest bias would follow a band-to-band tunneling current trend. β would not be well constrained since the trap-to-band tunneling effect was more noticeable in these arrays at the largest applied bias, where it is important to note that the large band-to-band tunneling currents did not allow for large enough well depths, where the median well depth for H1RG-20302 and 20303 are 243 and 306 mV respectively when 350 mV of bias was applied at 23 K. Therefore, in order to get an accurate estimate of β by obtaining several more points that are dominated by band-to-band tunneling currents, dark current and well depth data were obtained with 400 mV of applied bias at 23 K with an integration time of 167 ms, addressing only the 32 rows that were used to take warm-up data. Figure 6.10 shows the I-V curve for H1RG-20303 at a temperature of 23 K used to estimate β.

Figure 6.11 shows the β parameter distribution for all LW15 arrays. To show the improvement in array design, the median β parameter from H1RG-18509 was

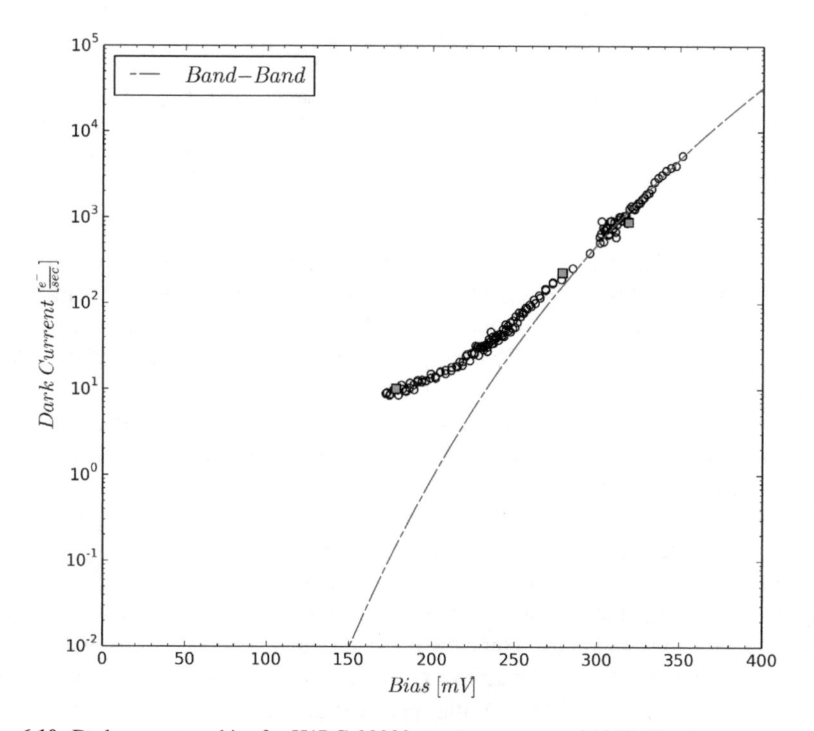

Fig. 6.10 Dark current *vs.* bias for H1RG-20303 at a temperature of 23 K. The β parameter was fitted using the I-V curve obtained from the 250 and 350 mV SUTR curves used to measure dark current and well depth. Additionally, I-V data obtained from 400 mV applied bias dark SUTR was used to add several more points that lie along exclusively on the band-to-band curve. The initial dark current and well depth for this pixel with 150, 250, and 350 mV were also added as cyan squares. It will be shown in Fig. 6.14 that the lower bias data not affected by band-to-band tunneling is affected by a combination of light leak, G-R, and trap-to-band tunneling

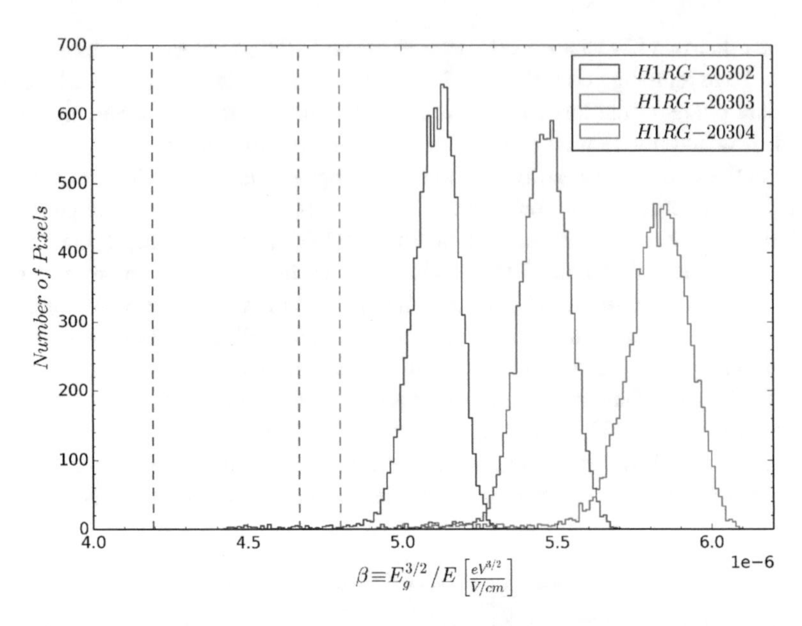

Fig. 6.11 Fitted β value distribution for the three LW15 arrays, with an actual reverse bias of 300 mV, to operable pixels in the central 32 rows (for which we have warm-up data) and 400 columns of the array. The vertical dashed lines correspond to the median β value of H1RG-18509 if the wavelength of the array were extrapolated to those of the three LW15 arrays

extrapolated to the three cutoff wavelengths of the LW15 arrays, and is shown as the vertical dashed lines in Fig. 6.11. The increased β parameter decreases tunneling currents. The β parameter distribution for H1RG-20302 is not a direct comparison with those of the other two LW15 arrays since the cutoff wavelength of the device is relatively larger, and therefore increases the β parameter.

The next step is to fit the thermally generated dark currents (diffusion and G-R) to the higher temperature I-T data obtained with an applied reverse bias of 150 mV. If the fitted band-to-band tunneling, diffusion, and G-R do not compensate for all of the dark current behavior over the entire I-T and I-V curves, then a light leak and trap-to-band tunneling current are fitted.

To show the uniformity of band-to-band tunneling, similar to the LW13 arrays, Fig. 6.12 shows the individual I-V curves of 36 operable pixels (randomly selected from the 32 $\times$ 400 region for which the parameter β was fitted). The individual I-V curves were obtained from the 250 and 350 mV dark SUTR data at a temperature of 23 K, along with the median initial dark current and well depth from 50, 150, 250, and 350 mV dark SUTR data for the 36 pixels. The median of individual I-T curves and the initial dark current at stable temperatures were used to fit the thermal dark currents. The error bars on the stable temperature data (or initial dark current and bias on I-V curve) corresponds to one standard deviation from the mean among the values from the 36 pixels. Table 6.6 contains the optimized parameters to fit the dark current behavior of the median of 36 pixels in all three LW15 arrays.

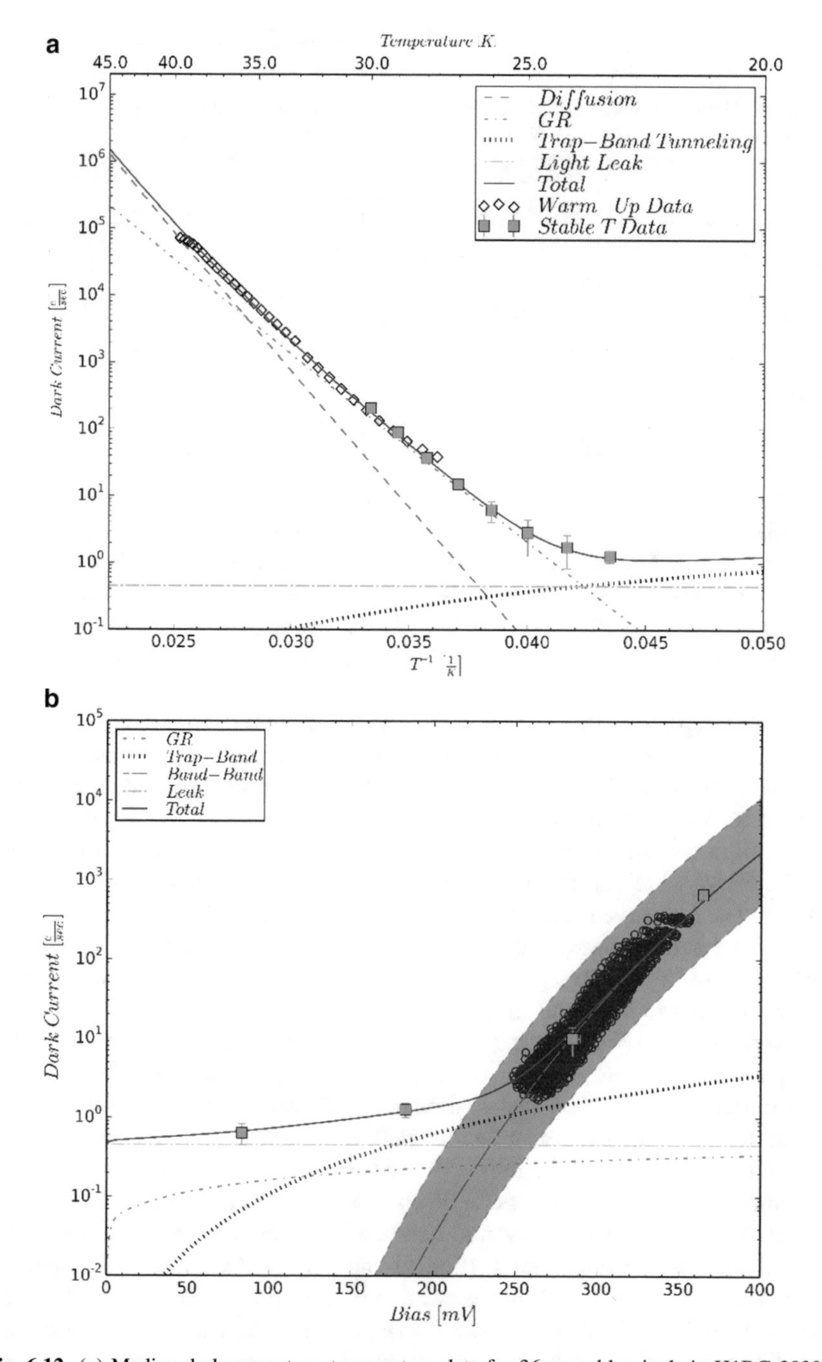

Fig. 6.12 (**a**) Median dark current *vs.* temperature data for 36 operable pixels in H1RG-20304. The median detector bias of 183 mV, corresponding to the 23 K stable data point, was used to fit the dark current models. The error bars on the stable temperature data correspond to ± one standard deviation of the mean in the dark current values for the averaged pixels. (**b**) Individual dark current *vs.* bias curves for the same 36 pixels obtained from the 250 and 350 mV SUTR curves obtained to measure dark current and well depth for this array at a temperature of 23 K. In addition to the individual I-V curves, the median initial dark current and well depth of SUTR curves at 150, 250, and 350 mV were added (cyan squares), where the error bars correspond to ± one standard deviation of the mean in the measurements for the pixels

Table 6.6 Optimized dark current parameters to model the median dark current behavior of a collection of pixels for all three LW15 arrays shown in Figs. 6.12, 6.13, and 6.14

	H1RG-20302 Fig. 6.13	H1RG-20303 Fig. 6.14	H1RG-20304 Fig. 6.12
$\beta\ \left[\frac{eV^{3/2}}{V/cm}\right]$	5.15×10^{-6}	5.4×10^{-6}	5.84×10^{-6}
$\tau_h\ [s]$	6.2×10^{-7}	2×10^{-7}	3.8×10^{-7}
$\tau_{GR}\ [s]$	6.2×10^{-5}	4.7×10^{-5}	1.1×10^{-5}
$E_{t_{gr}}\ [eV]$	0.041	0.043	0.052
$E_t\ [eV]$	0.063	0.068	0.062
$n_t\ \left[cm^{-3}\right]$	2.9	3.7	10
Light leak $\left[e^-/s\right]$	2.9	1.48	0.45

H1RG-20304 had the best performance from all three LW15 arrays since it has the shortest cutoff wavelength. Figure 6.12 shows the dark current behavior as a function of temperature and bias, where at biases greater than ~250 mV, band-to-band affects most pixels very uniformly. At lower biases, the dark current is composed of a combination of trap-to-band tunneling, G-R, and a small light leak ($<1\ e^-/s$).

A single trap density model of trap-to-band tunneling was fitted to the initial dark current and well depth values (square data points) to account for the bias dependent dark current behavior between ~80–250 mV that cannot be explained by light leak, G-R, or accounted for by band-to-band tunneling alone. From the relatively small error bars in the median of the data from the 36 pixels, there was no indication of a highly variable trap-to-band tunneling among these operable pixels at 50 and 150 mV. The I-V curve from the 50 and 150 mV dark SUTR is not plotted since the SUTR curve at those biases did not debias considerably. Unlike the LW13 devices, several sets of 36 operable pixels that were analyzed for all three LW15 devices showed this trap-to-band tunneling effect at lower biases.

The other two LW15 devices did not perform as well compared to H1RG-20304 for two different reasons—20302 had the longest cutoff wavelength of the three devices, while 20303 had a similar cutoff wavelength as 20304, but had slightly lower β parameter distribution: Both reasons lead to an increase in band-to-band tunneling currents. Figures 6.13 and 6.14 show the dark current behavior as functions of temperature and bias for 36 operable pixels in H1RG-20302 and 20303 respectively. The higher temperature dark current data is as expected dominated by G-R and/or diffusion. The portion of the I-V curves constructed from the 400 mV applied bias SUTR curves were very uniformly dominated by band-to-band tunneling. For these two devices the single trap density model of trap-to-band tunneling was used to compensate for the bias dependent dark current that was not accounted for with G-R alone or a light leak.

Though the single trap model would work to model the behavior of many of the individual plotted pixels for both devices, a few of the pixels do appear to show

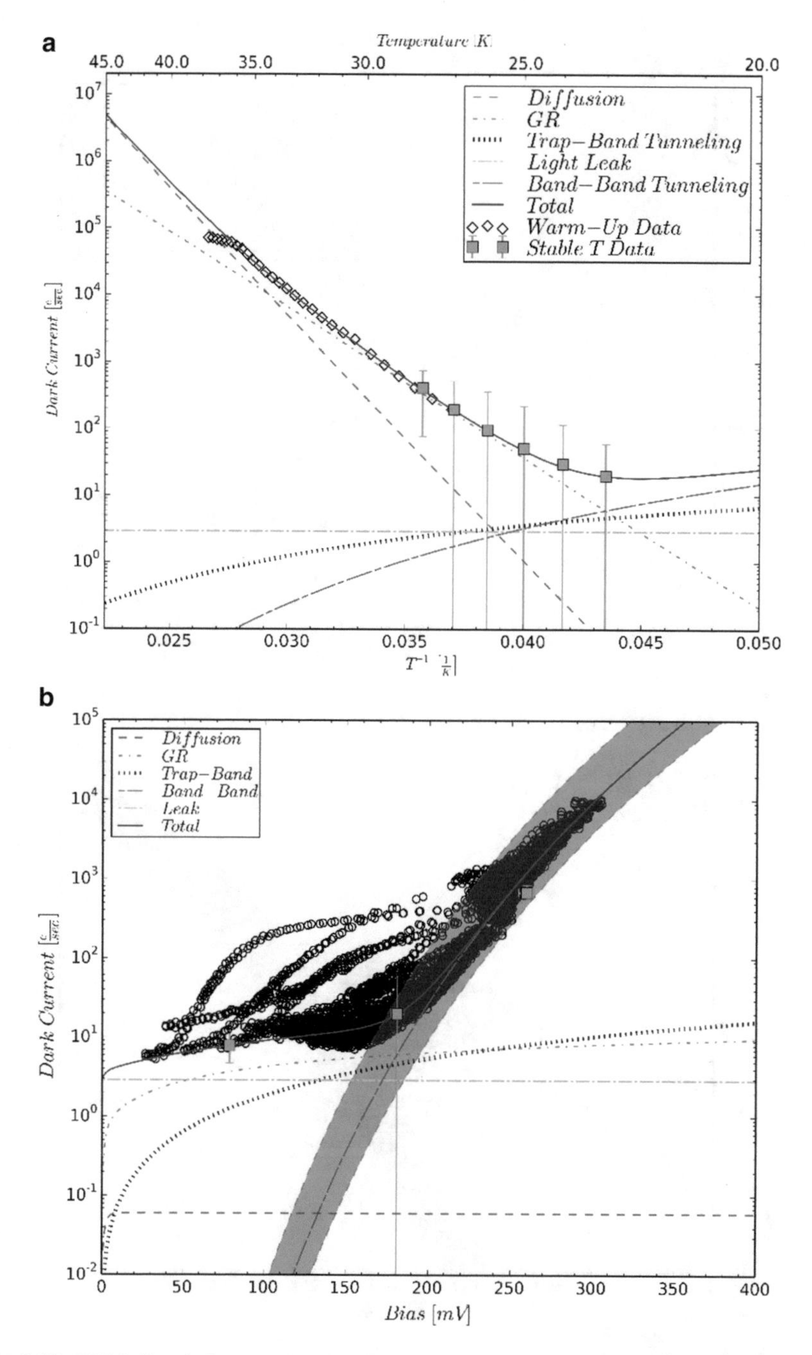

Fig. 6.13 (**a**) Median dark current *vs.* temperature data for 36 operable pixels in H1RG-20302. The median detector bias of 181 mV, corresponding to the 23 K stable data point, was used to fit the dark current models. The error bars on the median stable temperature data correspond to ± one standard deviation of the mean in the dark current values for the pixels. (**b**) Individual dark current *vs.* bias curves for the same 36 pixels obtained from the 250, 350, and 400 mV SUTR curves obtained to measure dark current and well depth for this array at a temperature of 23 K. In addition to the individual I-V curves, the median initial dark current and well depth of SUTR curves at 150, 250, and 350 mV were added, where the error bars correspond to ± one standard deviation of the mean in the measurements for the pixels

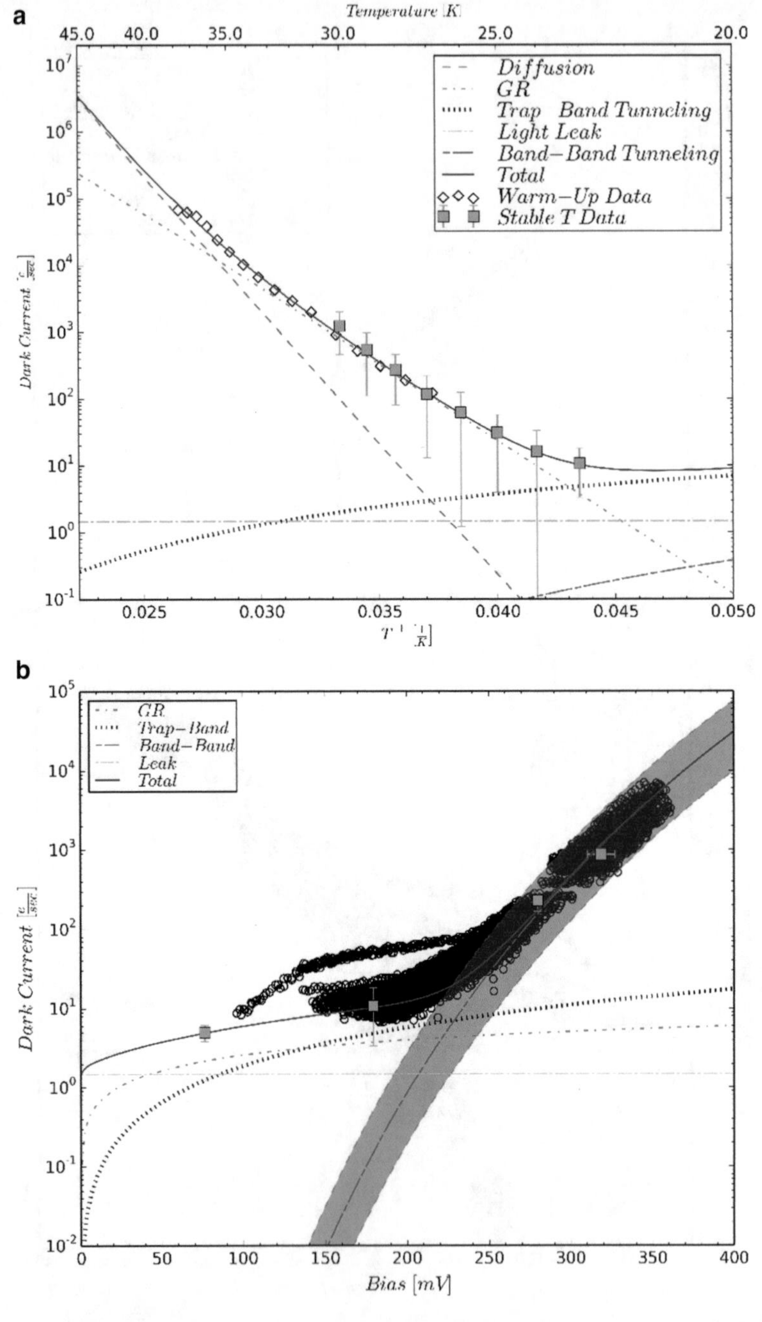

Fig. 6.14 (**a**) Median dark current *vs.* temperature data for 36 operable pixels in H1RG-20303. The median detector bias of 179 mV, corresponding to the 23 K stable data point, was used to fit the dark current models. The error bars on the median stable temperature data correspond to ± one standard deviation of the mean in the dark current values for the pixels. (**b**) Individual dark current *vs.* bias curves for the same 36 pixels obtained from the 250, 350, and 400 mV SUTR curves obtained to measure dark current and well depth for this array at a temperature of 23 K. In addition to the individual I-V curves, the median initial dark current and well depth of SUTR curves at 150, 250, and 350 mV were added, where the error bars correspond to ± one standard deviation of the mean in the measurements for the pixels

the soft breakdown that was seen in inoperable pixels in the LW13 devices, and the LW10 devices studied by Wu [17] and Bacon [20].

It is also important to note that the light leak and trap-to-band tunneling that were fitted here may not have a unique solution. Each initial guess given to the least squares optimizing function will result in different fitted parameters that fit the data well. Regardless of the fitted parameters for these two components, there is a dark current component that is bias dependent, which can be explained by trap-to-band tunneling. The light leak would not vary with bias to explain this bias-dependent behavior. While G-R does have a small bias dependence, it is not enough to compensate for the observed behavior. A combination of the two, light leak + G-R was also fitted to the data, where most of the results showed the total fitted dark current overestimating the dark current at the lowest biases, and missing the upturn of the dark current in the 200–250 mV range in H1RG-20303 and H1RG-20304, and in the 150–200 mV range in H1RG-20302 that cannot be explained by band-to-band tunneling alone. Figures 6.15 and 6.16 show the dark current model fit to data for an individual pixel in H1RG-20302 to show the bias dependence on dark current at lower bias that cannot be modeled well with only G-R and a light leak, and a dark current source with a larger bias dependence like trap-to-band tunneling is needed. The parameters used to fit the data for these two cases are shown in Table 6.7.

6.3 Phase II Summary

The characterization of the LW15 arrays showed that at a temperature of 28 K and an applied reverse bias of 50 mV, 96.3% of the pixels in H1RG-20304 have dark currents less than 200 e^-/s and well depths of at least 18 ke^-, where the percentage increases at lower temperatures (98.3% at 23 K). At larger applied biases, needed for larger well depths, H1RG-20304 is able to achieve ~80 % of pixels with dark currents less than 200 e^-/s and at least 60 ke^- (250 mV of applied reverse bias) up to a temperature of 27 K. Phase I of this project was crucial to determine the best array design to mitigate tunneling currents. As discussed in Sect. 6.1.3, the measured dark current in H1RG-20304, at an applied reverse bias of 250 mV, is nearly five orders of magnitude smaller than would be expected if this array had the nominal array design of the NEOCam LW10 devices.

Similar effects observed in the LW13 devices were also observed in the LW15 arrays, namely the dominant dark current behavior of band-to-band tunneling at the largest biases. The uniformity of this component of dark current is demonstrated by the uniform distribution of the β parameter. A similar analysis of trap-to-band tunneling to show the increased presence of this component among well behaved pixels would be difficult since this dark current component has several more parameters to fit, and the initial guess would have to be tailored for individual pixels with a substantially different behavior. Additionally, for the well behaved pixels over the entire bias range for which dark current models are fitted, trap-to-

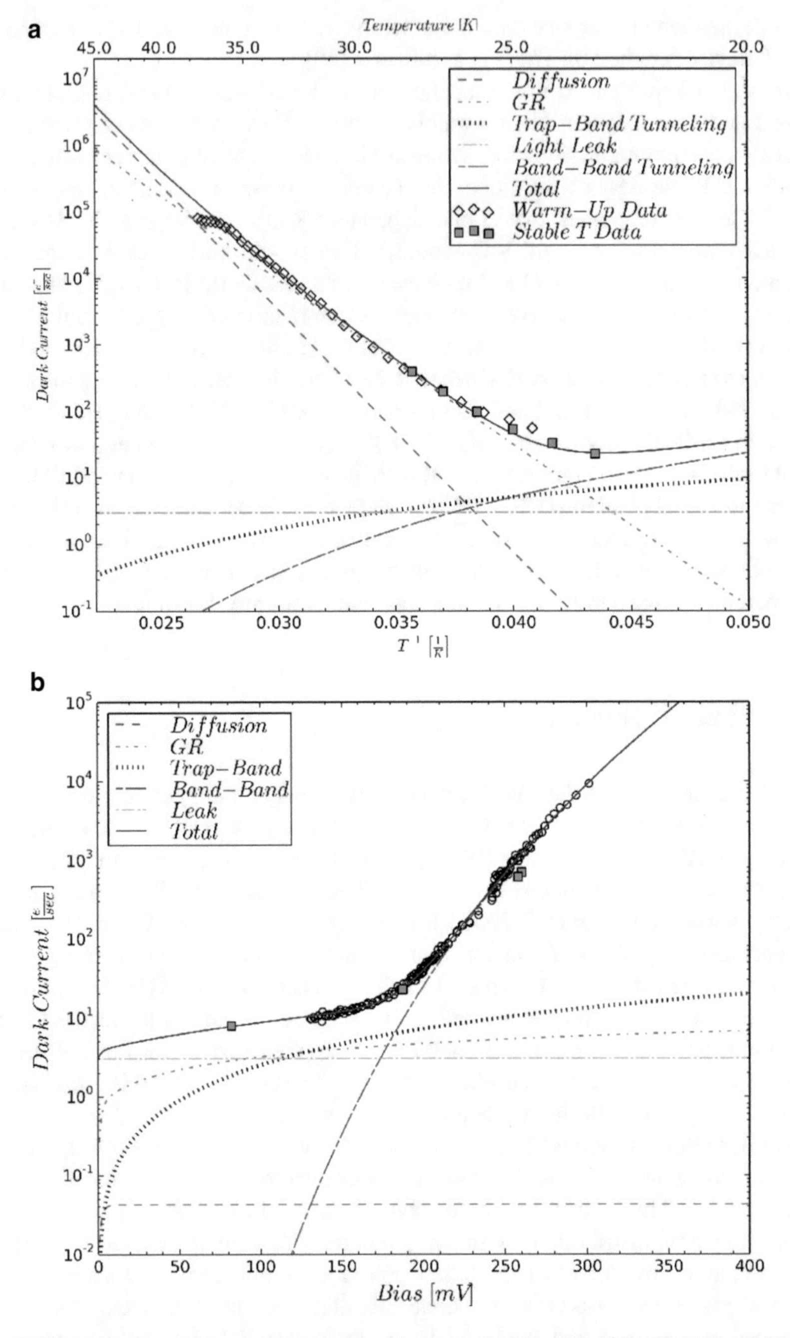

Fig. 6.15 (**a**) Dark current *vs.* temperature with an applied bias of 150 mV (detector bias of 187 mV) for pixel [516, 559] in H1RG-20302. (**b**) Dark current *vs.* bias (at 23 K). The cyan square data points are the initial dark current and detector bias for dark SUTR curves obtained with applied biases of 50, 150, 250, and 350 mV. 400 mV dark current and well depth data was obtained for only 32 rows of pixels with an integration time of 167 ms, allowing us to measure dark currents at higher biases before large dark currents debiased the pixels. The 167 ms data corresponds to the open circle data above ~250 mV

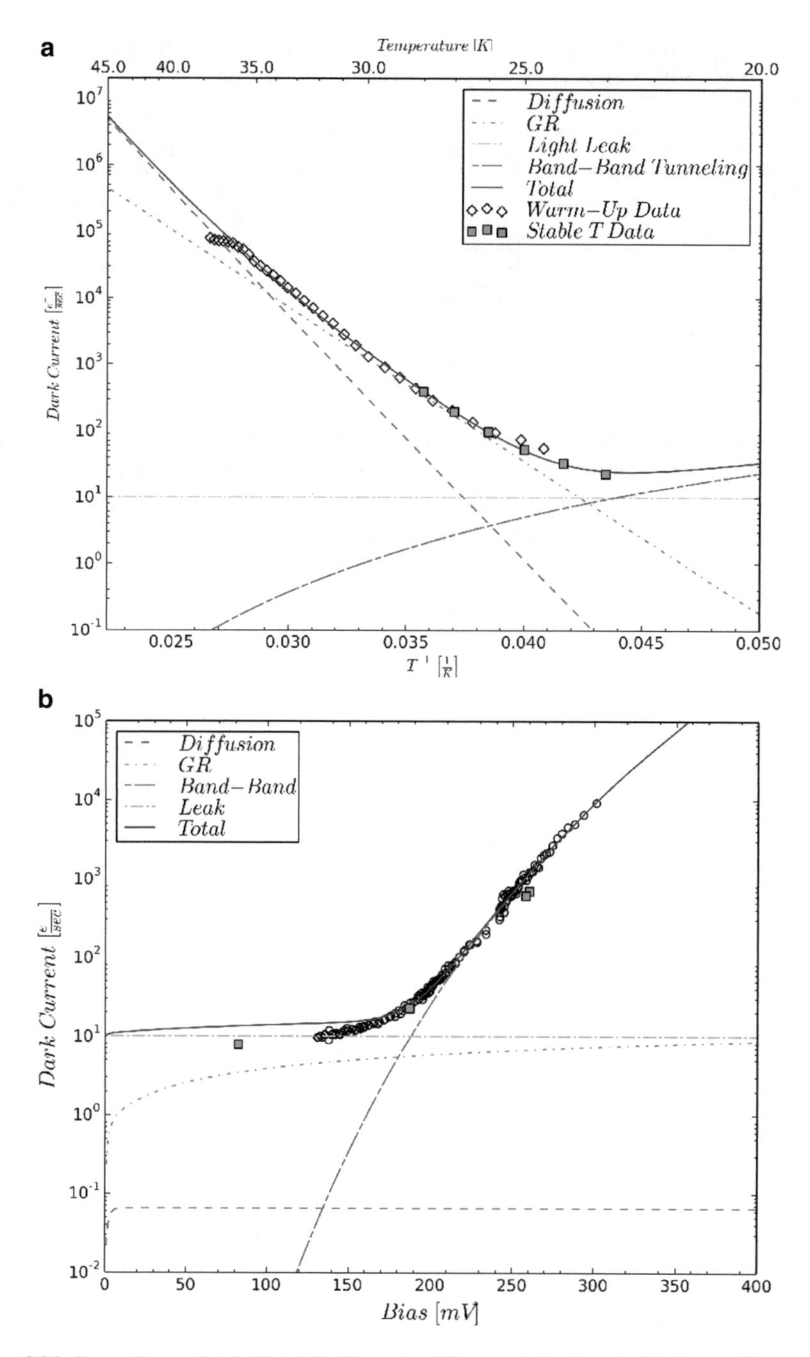

Fig. 6.16 Same data as that in Fig. 6.15. The only difference is that trap-to-band tunneling model is not fitted to the data. This shows that G-R and a light leak alone do not describe the lower bias behavior of the data well, and a dark current component with a larger bias dependence than a light leak and G-R at lower biases is needed to describe the low bias behavior

Table 6.7 Optimized dark current parameters to model the dark current behavior in pixel [516, 559] in H1RG-20302 with and without trap-to-band tunneling currents

	Figure 6.15	Figure 6.16
$\beta\ \left[\frac{eV^{3/2}}{V/cm}\right]$	5.16×10^{-6}	5.16×10^{-6}
$\tau_h\ [s]$	8.8×10^{-7}	5.7×10^{-7}
$\tau_{GR}\ [s]$	1.2×10^{-5}	4.4×10^{-5}
$E_{t_{gr}}\ [eV]$	0.045	0.042
$E_t\ [eV]$	0.064	–
$n_t\ \left[cm^{-3}\right]$	3	–
Light leak $\left[e^-/s\right]$	3	10

band tunneling currents are not the clear prevalent component, and would require the fitting of other components like G-R and the light leak in the test dewar.

Even with the apparent increase in trap-to-band tunneling currents among the well behaved pixels, the main takeaway from the results presented here is the successful improvement in array design by TIS to further reduce tunneling currents, driven by the results from the LW13 phase arrays.

Chapter 7
Conclusions and Future Work

The goal of this project has been to determine if HgCdTe detector arrays are a viable option for future passively cooled space astronomy missions covering the long wavelength range between 10 and 15 μm, replacing Si:As devices which have to be operated at a temperature of 6–8 K. The cutoff wavelength was chosen specifically for future missions that intend to study the atmospheres of exoplanets, where all three rocky planets in our solar system within the habitable zone and that have an atmosphere, all show a strong and broad spectral absorption feature due to the presence of CO_2 in their atmosphere. This feature can be used to identify rocky planets in other solar systems within their respective habitable zones.

Developing HgCdTe devices that could potentially work at passively cooled temperatures (Spitzer Space Telescope's focal plane array equilibrated at 27.5 K after cryogens were exhausted) with relatively low dark currents would benefit these future space missions since the lifetime of these missions would no longer be limited by the lifetime of cryogens or cryo-cooler malfunction. It has been shown that HgCdTe detector arrays with a wavelength of 10 μm can operate at temperatures of up to 40 K with a median dark current of less than 1 e^-/s, but extending the cutoff wavelength from 10 to 15 μm would prove difficult, mainly due to the expected increase in dark current due to the relatively smaller band gap of these devices. For the first phase of this project, the decision was made to first develop HgCdTe detector arrays with a cutoff wavelength of 13 μm to identify any unforeseen effects associated with increasing the cutoff wavelength of the devices, and determine the best solution to mitigate these effects to guarantee the success of the second phase of the project to reach the final goal of 15 μm cutoff wavelength.

For the first phase of the project, TIS delivered four LW13 devices to UR for characterization, where two of the arrays (H1RG-18367 and 18508) had the same standard design as the NEOCam LW10 devices and with an extended cutoff wavelength. The other two devices, H1RG-18369 and 18509, had two different experimental designs to mitigate tunneling dark currents. In Chap. 5, the characterization results of four ~13 μm cutoff wavelength devices are presented.

M. Cabrera, *Development of 15 Micron Cutoff Wavelength HgCdTe Detector Arrays for Astronomy*, Springer Theses, https://doi.org/10.1007/978-3-030-54241-2_7

It was shown that the median dark current of three of these devices had a median dark current of less than 1 e^-/s at a temperature of 28 K and 150 mV of applied bias.[1] At 28 K and 150 mV, more than 90% of the pixels in all four LW13 megapixel arrays showed dark currents <200 e^-/s, and well depths of at least >37 ke^-. Larger applied biases would be required to achieve larger well depths with the HXRG multiplexers, where it was shown in Chap. 5 that larger applied biases (>200 mV) will result in an exponentially increasing dark current for three of the four devices, where the median dark currents for these three devices were between 379–780 e^-/s at a temperature of 28 K and an applied bias of 350 mV. This increasing dark current is due to tunneling dark currents, where we were able to show that band-to-band tunneling was the main source of tunneling currents at a bias of 350 mV (largest bias applied in the characterization of these devices), and affected the majority of pixels uniformly.

The successful experimental design of LW13 detector array H1RG-18509 mitigated these tunneling currents, where at a temperature of 28 K and an applied bias of 350 mV, the median dark current was 1.8 e^-/s. This device has an operability of 90.8% up to a temperature of 35 K and applied bias of 350 mV,[2] whereas the other three devices had operabilities of ~0% due to the large tunneling currents at this bias.

In addition to future space missions, ground based observatories can also benefit from these devices since the N-band wavelength range can be covered with these LW13 devices, and can be operated at temperatures that can be achieved with a cry-cooler, eliminating the need for cryogens. Colleagues at the University of Michigan and the University of Arizona have begun studying the possibility of using these LW13 devices on ground-based instruments, only bonded to a CTIA mux to achieve larger well depths (larger background from ground compared to space), with a constant small bias while signal is integrated. The small applied bias required by the CTIA mux would eliminate any potential contribution of tunneling currents if they were present in the devices. The CTIA multiplexer would not be ideal for space-based missions since these devices have larger power dissipation than the HXRG multiplexers, which can increase the operating temperature of the devices for passively cooled missions, or can increase the boil rate of on-board cryogens. However, missions like the proposed Origins Space Telescope[50] include abundant cryo-coolers, which would enable such CTIA muxes.

For the second phase of this project, the array design of LW13 device H1RG-18509 was selected for the 15 μm cutoff wavelength goal devices since tunneling currents were expected to increase with the longer cutoff wavelength. The three LW15 devices that were delivered to UR had cutoff wavelengths of 15.2, 15.5, and 16.7 μm (at a temperature of 30 K). It was shown in Chap. 6 that the dark

[1] The fourth array (H1RG-18367) exhibited a multiplexer glow, increasing the current measured in the dark to 10 e^-/s.

[2] Operability requirements for this device with an applied bias of 350 mV consists of dark currents <200 e^-/s and well depths >75 ke^-.

current in all three devices at a temperature of 23 K and applied bias of 350 mV, band-to-band tunneling was the dominating dark current component. The measured dark current that is dominated by band-to-band tunneling in all three LW15 devices was lower than what would be expected from LW13 device H1RG-18509 if the cutoff wavelength were extended to the respective wavelengths of the LW15 devices, showing an improvement in the array design by TIS to mitigate tunneling dark currents.

Although there was a an increased contribution from what appears to be trap-to-band tunneling at lower biases among well behaved pixels (not observed in operable pixels in the LW13 devices), and an increase in G-R and diffusion current, the most important result is that the dominating dark current component at larger bias is band-to-band tunneling, where further improvement in array design by TIS could continue to decrease this dark current mechanism.

LW15 array H1RG-20304 has a cutoff wavelength closest to the goal of this project and was the best performing LW15 device, where at a temperature of 28 K and an applied bias of 250 mV, 80% of the pixels have dark currents below 200 e^{-}/s and well depths greater than 54 ke^{-} (median dark current and well depth of 58 e^{-}/s and 67 ke^{-} respectively).

Future work will be focused on determining the image quality that can be achieved with these devices by measuring the modulation transfer function (MTF), and assess the impact of the brighter-fatter effect on image quality. The brighter-fatter effect can increase the MTF (related to the point spread function) since the electron-hole pair produced in a saturated pixel can migrate to neighboring pixels, contributing to the blurring of a point source. The image quality of the LW10 arrays for NEOCam have shown good image quality[51], and little change is expected in the LW13 or LW15 detector arrays.

The future use of this technology in space missions is very promising, where the characterization of these devices has demonstrated the ability of this technology to operate at temperatures that can be attained through passive cooling in space. The comparison of the measured dark current against the theoretical model for various dark current mechanisms present as a function of temperature and bias will provide the information needed to focus future array improvement efforts on a distinct dark current component and/or determine the optimum operating conditions to meet the specific dark current requirements for future space or ground-based astronomy applications.

Bibliography

1. Werner, M.W., Roellig, T.L., Low, F.J., Rieke, G.H., Rieke, M., Hoffmann, W.F., Young, E., Houck, J.R., Brandl, B., Fazio, G.G., Hora, J.L., Gehrz, R.D., Helou, G., Soifer, B.T., Stauffer, J., Keene, J., Eisenhardt, P., Gallagher, D., Gautier, T.N., Irace, W., Lawrence, C.R., Simmons, L., Cleve, J.E.V., Jura, M., Wright, E.L., Cruikshank, D.P.: The Spitzer space telescope mission. Astrophys. J. Suppl. Ser. **154**(1), 1–9 (2004)
2. Gehrz, R.D., Roellig, T.L., Werner, M.W., Fazio, G.G., Houck, J.R., Low, F.J., Rieke, G.H., Soifer, B.T., Levine, D.A., Romana, E.A., The NASA Spitzer space telescope. Rev. Sci. Instrum. **78**(1), 011302 (2007)
3. Mainzer, A., Larsen, M., Stapelbroek, M.G., Hogue, H., Garnett, J., Zandian, M., Mattson, R., Masterjohn, S., Livingston, J., Lingner, N., Alster, N., Ressler, M., Masci, F.: Characterization of flight detector arrays for the wide-field infrared survey explorer. Proc. SPIE **7021**, 12 (2008)
4. Kaltenegger, L.: How to characterize habitable worlds and signs of life. Annu. Rev. Astron. Astrophys. **55**(1), 433–485 (2017)
5. Kittel, C.: Introduction to Solid State Physics. Wiley, New York (1996)
6. Ashcroft, N.W., Mermin, N.D.: Solid State Physics. Holt, Rinehart and Winston, New York (1976)
7. Hansen, G.L., Schmit, J.L.: Calculation of intrinsic carrier concentration in $Hg_{1-x}Cd_xTe$. J. Appl. Phys. **54**(3), 1639–1640 (1983)
8. Overhof, H.: A model calculation for the energy bands in the Hg1-xCdxTe mixed crystal system. Phys. Status Solidi B **45**(1), 315–321 (1971)
9. Hansen, G.L., Schmit, J.L., Casselman, T.N.: Energy gap versus alloy composition and temperature in hg1-xcdxte. J. Appl. Phys. **53**(10), 7099–7101 (1982)
10. Sze, S.M.: Physics of Semiconductor Devices. Wiley, New York (1981)
11. Moll, J.: Physics of Semiconductors. McGraw-Hill Physical and Quantum Electronics Series. McGraw-Hill, New York (1964)
12. Reine, M., Sood, A., Tredwell, T.: Photovoltaic infrared detectors. In: Willardson, R., Beer, A.C. (eds.) Mercury Cadmium Telluride, Semiconductors and Semimetals, vol. 18, Chap 6, pp. 201–311. Elsevier, Amsterdam (1981)
13. Kinch, M.A.: State-of-the-Art Infrared Detector Technology. SPIE Press, Bellingham (2014)
14. Fowler, A.M., Gatley, I.: Demonstration of an algorithm for read-noise reduction in infrared arrays. Astrophys. J. Lett. **353**, L33 (1990)
15. Rieke, G.H.: Detection of Light: From the Ultraviolet to the Submillimeter. Cambridge University Press, Cambridge (2003)
16. Forrest, W.J.: Private communication (2014)

M. Cabrera, *Development of 15 Micron Cutoff Wavelength HgCdTe Detector Arrays for Astronomy*, Springer Theses, https://doi.org/10.1007/978-3-030-54241-2

17. Wu, J.: Development of infrared detectors for space astronomy. PhD thesis, University of Rochester (1997)
18. Bailey, R.B., Arias, J.M., McLevige, W.V., Pasko, J.G., yi Chen, A.C., Cabelli, C.A., Kozlowski, L.J., Vural, K., Wu, J., Forrest, W.J., Pipher, J.L.: Prospects for large-format IR astronomy FPAs using MBE-grown HgCdTe detectors with cutoff wavelength $> 4\ \mu$m. Proc. SPIE **3354**, 10 (1998)
19. Bacon, C.M., McMurtry, C.W., Pipher, J.L., Forrest, W.J., Garnett, J.D., Lee, D., Edwall, D.D.: Further characterization of Rockwell scientific LWIR HgCdTe detector arrays. Proc. SPIE **5563**, 11 (2004)
20. Bacon, C.M.: Development of long wave infrared detectors for space astronomy. PhD thesis, University of Rochester (2006)
21. Bacon, C.M., McMurtry, C.W., Pipher, J.L., Mainzer, A., Forrest, W.: Effect of dislocations on dark current in LWIR HgCdTe photodiodes. Proc. SPIE **7742**, 9 (2010)
22. McMurtry, C., Lee, D.L., Beletic, J., Chen, C.-Y.A., Demers, R.T., Dorn, M., Edwall, D.D., Fazar, C.M., Forrest, W.J., Liu, F., Mainzer, A.K., Pipher, J.L., Yulius, A.: Development of sensitive long-wave infrared detector arrays for passively cooled space missions. Opt. Eng. **52**, 091804 (2013)
23. Smith, E.C., Rauscher, B.J., Alexander, D., Clemons, B.L., Engler, C., Garrison, M.B., Hill, R.J., Johnson, T., Lindler, D.J., Manthripragada, S.s., Marshall, C., Mott, B., Parr, T.M., Roher, W.D., Shakoorzadeh, K.B., Schnurr, R., Waczynski, A., Wen, Y., Wilson, D., Loose, M., Bagnasco, G., Böker, T., Marchi, G.D., Ferruit, P., Jakobsen, P., Strada, P.: JWST near infrared detectors: latest test results. Proc. SPIE **7419**, 10 (2009)
24. Dorn, M., McMurtry, C., Pipher, J., Forrest, W., Cabrera, M., Wong, A., Mainzer, A.K., Lee, D., Pan, J.: A monolithic 2k x 2k LWIR HgCdTe detector array for passively cooled space missions. Proc. SPIE **10709**, 1070907 (2018)
25. Dorn, M.L., Pipher, J.L., McMurtry, C.W., Hartman, S., Mainzer, A., McKelvey, M., McMurray, R., Chevara, D., Rosser, J.: Proton irradiation results for long-wave HgCdTe infrared detector arrays for Near-Earth Object Camera. J. Astron. Telesc. Instrum. Syst. **2**(3), 11 (2016)
26. McMurtry, C.W., Dorn, M., Cabrera, M.S., Pipher, J.L., Forrest, W.J., Mainzer, A.K., Wong, A.: Candidate 10 micron HgCdTe Arrays for the NEOCam Space Mission. Proc. SPIE **9915**, 8 (2016)
27. Carmody, M., Lee, D., Zandian, M., Phillips, J., Arias, J.: Threading and misfit-dislocation motion in molecular-beam epitaxy-grown HgCdTe epilayers. J. Electron. Mater. **32**(7), 710–716 (2003)
28. Sah, C.T., Noyce, R.N., Shockley, W.: Carrier generation and recombination in P-N junctions and P-N junction characteristics. Proc. IRE **45**(9), 1228–1243 (1957)
29. Martinka, M., Almeida, L.A., Benson, J.D., Dinan, J.H.: Characterization of cross-hatch morphology of MBE (211) HgCdTe. J. Electron. Mater. **30**, 632–636 (2001)
30. Chang, Y., Becker, C., Grein, C., Zhao, J., Fulk, C., Casselman, T., Kiran, R., Wang, X., Robinson, E., An, S., Mallick, S., Sivananthan, S., Aoki, T., Wang, C., Smith, D., Velicu, S., Zhao, J., Crocco, J., Chen, Y., Brill, G., Wijewarnasuriya, P., Dhar, N., Sporken, R., Nathan, V.: Surface morphology and defect formation mechanisms for HgCdTe (211)B grown by molecular beam epitaxy. J. Electron. Mater. **37**, 1171–1183 (2008)
31. Garnett, J.D., Forrest, W.J.: Multiply sampled read-limited and background-limited noise performance. Proc. SPIE **1946**, 395–404 (1993)
32. Rauscher, B.J., Fox, O., Ferruit, P., Hill, R.J., Waczynski, A., Wen, Y., Xia-Serafino, W., Mott, B., Alexander, D., Brambora, C.K., Derro, R., Engler, C., Garrison, M.B., Johnson, T., Manthripragada, S.S., Marsh, J.M., Marshall, C., Martineau, R.J., Shakoorzadeh, K.B., Wilson, D., Roher, W.D., Smith, M., Cabelli, C., Garnett, J., Loose, M., Wong-Anglin, S., Zandian, M., Cheng, E., Ellis, T., Howe, B., Jurado, M., Lee, G., Nieznanski, J., Wallis, P., York, J., Regan, M.W., Hall, D.N.B., Hodapp, K.W., Böker, T., Marchi, G.D., Jakobsen, P., Strada, P.: Detectors for the James Webb Space Telescope near-infrared spectrograph. I. Readout mode, noise model, and calibration considerations. Publ. Astron. Soc. Pac. **119**(857), 768 (2007)

33. Mortara, L., Fowler, A.: Evaluations of charge-coupled device (CCD) performance for astronomical use. Proc. SPIE **0290**, 6 (1981)
34. Shockley, W., Read, W.T.: Statistics of the recombinations of holes and electrons. Phys. Rev. **87**, 835–842 (1952)
35. Kinch, M.: Metal-insulator-semiconductor infrared detectors. In: Willardson, R., Beer, A.C. (eds.) Mercury Cadmium Telluride, Semiconductors and Semimetals, vol. 18, Chap. 7, pp. 313–378. Elsevier, Amsterdam (1981)
36. Neudeck, P., Huang, W., Dudley, M.: Breakdown degradation associated with elementary screw dislocations in 4H-SiC P^+N junction rectifiers. Solid-State Electron. **42**(12), 2157–2164 (1998)
37. Neudeck, P.G., Huang, W., Dudley, M.: Study of bulk and elementary screw dislocation assisted reverse breakdown in low-voltage (<250 V) 4H-SiC p^+n junction diodes. I: DC properties. IEEE Trans. Electron Devices **46**(3), 478–484 (1999)
38. Ravi, K.V., Varker, C.J., Volk, C.E.: Electrically active stacking faults in silicon. J. Electrochem. Soc. **120**(4), 533–541 (1973)
39. Benson, J.D., Bubulac, L.O., Smith, P.J., Jacobs, R.N., Markunas, J.K., Jaime-Vasquez, M., Almeida, L.A., Stoltz, A.J., Wijewarnasuriya, P.S., Brill, G., Chen, Y., Lee, U., Vilela, M.F., Peterson, J., Johnson, S.M., Lofgreen, D.D., Rhiger, D., Patten, E.A., Goetz, P.M.: Characterization of dislocations in (112)B HgCdTe/CdTe/Si. J. Electron. Mater. **39**, 1080–1086 (2010)
40. Rauscher, B.J., Figer, D.F., Regan, M.W., Bergeron, L.E., Balleza, J.C., Barkhouser, R.H., Greene, G.R., Kim, S., McCandliss, S.R., Morse, E., Pelton, R., Reeves, T., Sharma, U., Stemniski, P., Stockman, H.S., Telewicz, M.: Ultralow-background operation of near-infrared detectors using reference pixels for NGST. Proc. SPIE **4850**, 962–970 (2003)
41. Moore, A.C., Ninkov, Z., Forrest, W.J.: Interpixel capacitance in non-destructive focal plane arrays. Proc. SPIE **5167**, 12 (2004)
42. Donlon, K., Ninkov, Z., Baum, S.: Signal dependence of inter-pixel capacitance in hybridized HgCdTe H2RG arrays for use in James Webb space telescope's NIRcam. Proc. SPIE **9915** (2016). https://doi.org/10.1117/12.2233200
43. Wu, J., Forrest, W.J., Pipher, J.L., Lum, N., Hoffman, A.: Development of infrared focal plane arrays for space. Rev. Sci. Instrum. **68**(9), 3566–3578 (1997)
44. Plazas, A., Shapiro, C., Smith, R., Rhodes, J., Huff, E.: Nonlinearity and pixel shifting effects in HXRG infrared detectors. J. Instrum. **12**, C04009 (2017)
45. Ayers, J.: Heteroepitaxy of Semiconductors: Theory, Growth, and Characterization. CRC Press, Boca Raton (2007)
46. Shapiro, C., Huff, E., Smith, R.: Intra-pixel response characterization of a HgCdTe near infrared detector with a pronounced crosshatch pattern. Proc. SPIE **10709**, 1070936 (2018)
47. Rauscher, B.J., Arendt, R.G., Fixsen, D.J., Greenhouse, M.A., Lander, M., Lindler, D., Loose, M., Moseley, S.H., Mott, D.B., Wen, Y., Wilson, D.V., Xenophontos, C.: Principal components analysis of a JWST NIRSpec detector subsystem. Proc. SPIE **8860**, 886005 (2013)
48. Fossum, E.R., Pain, B.: Infrared readout electronics for space-science sensors: state of the art and future directions. Proc. SPIE **2020** (1993). https://doi.org/10.1117/12.160549
49. Loose, M., Beletic, J., Garnett, J., Xu, M.: High-performance focal plane arrays based on the HAWAII-2RG/4RG and the SIDECAR ASIC. Proc. SPIE **6690**, 66900C (2007)
50. Leisawitz, D., Amatucci, E., Carter, R., DiPirro, M., Flores, A., Staguhn, J., Wu, C., Allen, L., Arenberg, J., Armus, L., Battersby, C., Bauer, J., Bell, R., Beltran, P., Benford, D., Bergin, E., Bradford, C.M., Bradley, D., Burgarella, D., Carey, S., Chi, D., Cooray, A., Corsetti, J., Beck, E.D., Denis, K., Dewell, L., East, M., Edgington, S., Ennico, K., Fantano, L., Feller, G., Folta, D., Fortney, J., Generie, J., Gerin, M., Granger, Z., Harpole, G., Harvey, K., Helmich, F., Hilliard, L., Howard, J., Jacoby, M., Jamil, A., Kataria, T., Knight, S., Knollenberg, P., Lightsey, P., Lipscy, S., Mamajek, E., Martins, G., Meixner, M., Melnick, G., Milam, S., Mooney, T., Moseley, S.H., Narayanan, D., Neff, S., Nguyen, T., Nordt, A., Olson, J., Padgett, D., Petach, M., Petro, S., Pohner, J., Pontoppidan, K., Pope, A., Ramspacher, D., Roellig, T., Sakon, I., Sandin, C., Sandstrom, K., Scott, D., Sheth, K., Steeves, J., Stevenson, K., Stokowski, L.,

Stoneking, E., Su, K., Tajdaran, K., Tompkins, S., Vieira, J., Webster, C., Wiedner, M., Wright, E.L., Zmuidzinas, J.: The origins space telescope: mission concept overview **10698**, 1069815 (2018)

51. Dorn, M.L.: Characterizing large format $10\mu m$ cutoff detector arrays for low background space applications. PhD thesis, University of Rochester (2019)

Biographical Sketch

The author was born in Irapuato, Guanajuato, Mexico to Maria Gabriela Cabrera Morales and Mario Cabrera Cabrera. He attended El Camino Community College and transferred to California State Polytechnic University, Pomona in 2011, where he graduated with a Bachelor of Science degree in Physics in 2014. He began his graduate studies in 2014 at the University of Rochester in the Physics and Astronomy department, where he was awarded a GAANN fellowship. For his doctoral research, he joined the UR infrared detector group led by Professor William Forrest and Professor Judith Pipher, focusing on the development of long wave infrared detector arrays for astronomy. In 2015 he earned a Master of Arts degree in Physics and his PhD in Physics and Astronomy in 2020. The author started working for Conceptual Analytics LLC as a Detector Engineer in 2019.

The following publications resulted from the author's doctoral research:

- Cabrera, M.S.; McMurtry, C.W.; Forrest, W.J.; Pipher, J.L.; Dorn, M.L.; Lee, D.; "Characterization of a 15 μm Cutoff HgCdTe Detector Array for Astronomy," Journal of Astronomical Telescopes, Instruments, and Systems **6**(1), 1–9 (2019).
- Cabrera, M. S., McMurtry, C. W., Dorn, M. L., Forrest, W. J., Pipher, J. L., and Lee, D., "Development of 13 μm Cutoff HgCdTe Detector Arrays for Astronomy," Journal of Astronomical Telescopes, Instruments, and Systems **5**(3), 1–18 (2019).
- McMurtry, C.W.; Cabrera, M.S.; Dorn, M.L.; Pipher, J.L.; Forrest, W.J.; "13 micron cutoff HgCdTe detector arrays for space and ground-based astronomy", Proc. SPIE **9915**, 99150E (2016).

M. Cabrera, *Development of 15 Micron Cutoff Wavelength HgCdTe Detector Arrays for Astronomy*, Springer Theses, https://doi.org/10.1007/978-3-030-54241-2

Printed in the United States
by Baker & Taylor Publisher Services